Hydrogeologic Framework and Hydrologic Budget Components of the Columbia Plateau Regional Aquifer System, Washington, Oregon, and Idaho

By S.C. Kahle, D.S. Morgan, W.B. Welch, D.M. Ely, S.R. Hinkle, J.J. Vaccaro, and L.L. Orzol

Groundwater Resources Program

Scientific Investigations Report 2011–5124

U.S. Department of the Interior
U.S. Geological Survey

U.S. Department of the Interior
KEN SALAZAR, Secretary

U.S. Geological Survey
Marcia K. McNutt, Director

U.S. Geological Survey, Reston, Virginia: 2011

For more information on the USGS—the Federal source for science about the Earth, its natural and living resources, natural hazards, and the environment, visit http://www.usgs.gov or call 1–888–ASK–USGS.

For an overview of USGS information products, including maps, imagery, and publications, visit http://www.usgs.gov/pubprod

To order this and other USGS information products, visit http://store.usgs.gov

Suggested citation:
Kahle, S.C., Morgan, D.S., Welch, W.B., Ely, D.M., Hinkle, S.R., Vaccaro, J.J., and Orzol, L.L., 2011, Hydrogeologic framework and hydrologic budget components of the Columbia Plateau Regional Aquifer System, Washington, Oregon, and Idaho: U.S. Geological Survey Scientific Investigations Report 2011–5124, 66 p.

Contents

Contents—Continued

Figures

Figures—Continued

Tables

Conversion Factors, Datums, and Abbreviations and Acronyms

Conversion Factors

Inch/Pound to SI

Multiply	By	To obtain
Length		
inch (in.)	2.54	centimeter (cm)
inch (in.)	25.4	millimeter (mm)
foot (ft)	0.3048	meter (m)
mile (mi)	1.609	kilometer (km)
Area		
acre	4,047	square meter (m²)
square foot (ft²)	0.09290	square meter (m²)
square mile (mi²)	2.590	square kilometer (km²)
Volume		
acre-foot	1233.4819	kiloliter (kl)
million gallons (Mgal)	3,785	cubic meter (m³)
Flow rate		
acre-foot per year (acre-ft/yr)	1,233	cubic meter per year (m³/yr)
foot per day (ft/d)	0.3048	meter per day (m/d)
cubic foot per second (ft³/s)	0.02832	cubic meter per second (m³/s)
cubic foot per day (ft³/d)	0.02832	cubic meter per day (m³/d)
gallon per minute (gal/min)	0.06309	liter per second (L/s)
gallon per day (gal/d)	0.003785	cubic meter per day (m³/d)
million gallons per day (Mgal/d)	3785.4118	cubic meters/day (m³/d)
inch per year (in/yr)	25.4	millimeter per year (mm/yr)
Hydraulic conductivity		
foot per day (ft/d)	0.3048	meter per day (m/d)
Transmissivity*		
cubic foot per second (ft³/s)	0.02832	cubic meter per second
cubic foot per second (ft³/s)	0.09290	square meter per day (m²/d)
foot squared per day (ft²/d)	0.09290	meter squared per day (m²/d)

Conversion Factors, Datums, and Abbreviations and Acronyms—Continued

Conversion Factors

SI to Inch/Pound

Multiply	By	To obtain
Length		
centimeter (cm)	0.3937	inch (in.)
millimeter (mm)	0.03937	inch (in.)
meter (m)	3.281	foot (ft)
kilometer (km)	0.6214	mile (mi)
Area		
square kilometer (km^2)	247.1	acre
Flux rate		
millimeter per day (mm/d)	0.03937	inch per day (in/d)

Temperature in degrees Celsius (°C) may be converted to degrees Fahrenheit (°F) as follows:

$$°F = (1.8 × °C) + 32$$

*Transmissivity: The standard unit for transmissivity is cubic foot per day per square foot times foot of aquifer thickness [(ft^3/d)/ft^2]ft. In this report, the mathematically reduced form, square foot per day (ft^2/d), is used for convenience.

Datums

Vertical coordinate information is referenced to the North American Vertical Datum of 1988 (NAVD 88).

Horizontal coordinate information is referenced to the North American Datum of 1983 (NAD 83).

Altitude, as used in this report, refers to distance above the vertical datum.

Conversion Factors, Datums, and Abbreviations and Acronyms—Continued

Abbreviations and Acronyms

‰	units of per mil
δ	delta
^{14}C	carbon-14
^{13}C	carbon-13
^{16}O	oxygen-16
^{18}O	oxygen-18
AVHRR	Advanced Very High Resolution Radiometer
CBP	Columbia Basin Project
CPRAS	Columbia Plateau Regional Aquifer System
CO_2	carbon dioxide
CRBG	Columbia River Basalt Group
DEM	digital elevation model
DIC	dissolved inorganic carbon
DR	direct runoff
EROS	Earth Resources Observation and Science Center
ET	evapotranspiration
ETa	actual evapotranspiration
ETr or ETp	reference (potential) evapotranspiration
GDAS	Global Data Assimilation System
GF	groundwater flux
IE	irrigation efficiency
IR	irrigation application
IT	irrigation target
K_h	horizontal hydraulic conductivity
K_v	vertical hydraulic conductivity
LOESS	local trend estimation method
LST	land-surface temperature
MAD	maximum allowable depletion
MAF	million acre-feet
MODIS	Moderate Resolution Imaging Spectroradiometer
NI	net infiltration
OWRD	Oregon Water Resources Department
PMC	percent modern carbon
PR	precipitation
PRISM	Parameter-elevation Regressions on Independent Slopes Model
PWS	public water supply
r^2	correlation coefficient
RMSE	root mean square error
SM	soil moisture
SMC	soil-moisture capacity
SOWAT	Soil Water Balance Model
STATSGO	State Soil Geographic Database
SSEB	Simplified Surface Energy Balance
SDWIS	USEPA's Safe Drinking Water Information System
USDA	U.S. Department of Agriculture
USEPA	U.S. Environmental Protection Agency
USGS	U.S. Geological Survey
WRTS	Water Rights Tracking System

x

Hydrogeologic Framework and Hydrologic Budget Components of the Columbia Plateau Regional Aquifer System, Washington, Oregon, and Idaho

By S.C. Kahle, D.S. Morgan, W.B. Welch, D.M. Ely, S.R. Hinkle, J.J. Vaccaro, and L.L. Orzol

Abstract

The Columbia Plateau Regional Aquifer System (CPRAS) covers an area of about 44,000 square miles in a structural and topographic basin within the drainage of the Columbia River in Washington, Oregon, and Idaho. The primary aquifers are basalts of the Columbia River Basalt Group (CRBG) and overlying sediment. Eighty percent of the groundwater use in the study area is for irrigation, in support of a $6 billion per year agricultural economy. Water-resources issues in the Columbia Plateau include competing agricultural, domestic, and environmental demands. Groundwater levels were measured in 470 wells in 1984 and 2009; water levels declined in 83 percent of the wells, and declines greater than 25 feet were measured in 29 percent of the wells.

Conceptually, the system is a series of productive basalt aquifers consisting of permeable interflow zones separated by less permeable flow interiors; in places, sedimentary aquifers overly the basalts. The aquifer system of the CPRAS includes seven hydrogeologic units—the overburden aquifer, three aquifer units in the permeable basalt rock, two confining units, and a basement confining unit. The overburden aquifer includes alluvial and colluvial valley-fill deposits; the three basalt units are the Saddle Mountains, Wanapum, and Grande Ronde Basalts and their intercalated sediments. The confining units are equivalent to the Saddle Mountains-Wanapum and Wanapum-Grande Ronde interbeds, referred to in this study as the Mabton and Vantage Interbeds, respectively. The basement confining unit, referred to as Older Bedrock, consists of pre-CRBG rocks that generally have much lower permeabilities than the basalts and are considered the base of the regional flow system. Based on specific-capacity data, median horizontal hydraulic conductivity (K_h) values for the overburden, basalt units, and bedrock are 161, 70, and 6 feet per day, respectively.

Analysis of oxygen isotopes in water and carbon isotopes in dissolved inorganic carbon from groundwater samples indicates that groundwater in the CPRAS ranges in age from modern (<50 years) to Pleistocene (>10,000 years). The oldest groundwater resides in deep, downgradient locations indicating that groundwater movement and replenishment in parts of this regional aquifer system have operated on long timescales under past natural conditions, which is consistent with the length and depth of long flow paths in the system.

The mean annual recharge from infiltration of precipitation for the 23-year period 1985–2007 was estimated to be 4.6 inches per year (14,980 cubic feet per second) using a polynomial regression equation based on annual precipitation and the results of recharge modeling done in the 1980s. A regional-scale hydrologic budget was developed using a monthly SOil WATer (SOWAT) Balance model to estimate irrigation-water demand, groundwater flux (recharge or discharge), direct runoff, and soil moisture within irrigated areas. Mean monthly irrigation throughout the study area peaks in July at 1.6 million acre-feet (MAF), of which 0.45 and 1.15 MAF are from groundwater and surface-water sources, respectively. Annual irrigation water use in the study area averaged 5.3 MAF during the period 1985–2007, with 1.4 MAF (or 26 percent) supplied from groundwater and 3.9 MAF supplied from surface water. Mean annual recharge from irrigation return flow in the study area was 4.2 MAF (1985–2007) with 2.1 MAF (50 percent) occurring within the predominately surface-water irrigated regions of the study area.

Annual groundwater-use estimates were made for public supply, self-supplied domestic, industrial, and other uses for the period 1984 through 2009. Public supply groundwater use within the study area increased from 200,600 acre-feet per year (acre-ft/yr) in 1984 to 269,100 acre-ft/yr in 2009. Domestic self-supplied groundwater use increased from 54,580 acre-ft/yr in 1984 to 71,160 acre-ft/yr in 2009. Industrial groundwater use decreased from 53,390 acre-ft/yr in 1984 to 43,930 acre-ft/yr in 2009.

Introduction

The Columbia Plateau Regional Aquifer System (CPRAS) covers approximately 44,000 mi² of northeastern Oregon, southeastern Washington, and western Idaho (fig. 1). The area supports a $6 billion per year agricultural industry, leading the Nation in production of apples and nine other commodities (State of Washington Office of Financial Management, 2007; U.S. Department of Agriculture, 2007). Groundwater availability in the aquifers of the area is a critical water-resource management issue because the water demand for agriculture, economic development, and ecological needs is high.

The primary aquifers of the CPRAS are basalts of the Columbia River Basalt Group (CRBG) and in places, overlying basin-fill sediments. Water-resources issues that have implications for current (2011) and future groundwater availability in the region include (1) widespread water-level declines associated with development of groundwater resources for irrigation and other uses, (2) reduction in base flow to rivers and associated effects on river temperature and water quality, and (3) current and anticipated effects of global climate change on recharge, base flow, and ultimately, groundwater availability.

The U.S. Geological Survey (USGS) Groundwater Resources Program began a study of the CPRAS in 2007 with the broad goals of (1) characterizing the hydrologic status of the system, (2) identifying trends in groundwater storage and use, and (3) quantifying groundwater availability. The study approach includes updating the regional hydrogeologic framework, documenting changes in the status of the system, quantifying the hydrologic budget, and developing a groundwater-flow simulation model for the system. The simulation model will be used to evaluate and test the conceptual model of the system and later to evaluate groundwater availability under alternative development and climate scenarios.

This report, which describes the hydrogeologic framework and selected groundwater-budget components of the CPRAS, along with three additional reports associated with this project (Kahle and others, 2009; Snyder and Haynes, 2010; and Burns and others, 2011), provides comprehensive information about the physical framework of the CPRAS based on historical and current investigations. The final phase of the study is using this information to develop a numerical groundwater-flow model to assess the groundwater availability of the CPRAS. The results of the overall investigation are intended to characterize the hydrologic status of the system, identify trends in groundwater storage and use, and quantify groundwater availability.

Purpose and Scope

The purpose of this report is to describe the hydrogeologic framework and selected hydrologic budget components of the CPRAS that will be used for the numerical groundwater-flow model. The description of the hydrogeologic framework is based on historical and recent investigations. The hydrologic budget components are based on new methods and new information compiled during this investigation. The scope of this report includes the regional geologic history; sediment and basalt stratigraphy; hydrogeologic units; hydraulic characteristics; and groundwater occurrence, movement, and approximate age. Additionally, selected hydrologic budget components of the CPRAS include estimates of recharge from infiltration of precipitation, irrigation water use and associated recharge, results from a monthly soil-water balance model, and estimates of non-irrigation water use.

Description of Study Area

The Columbia Plateau is a structural and topographic basin within the drainage of the Columbia River (fig. 1). It is bound on the west by the Cascade Range, on the east by the Rocky Mountains, and on the north by the Okanogan Highlands. Its southern boundary corresponds to the mapped extent of the CRBG. Major tributaries to the Columbia River include the Yakima, Spokane, Coeur D'Alene, Clearwater, Salmon, Imnaha, Grande Ronde, Snake, Tucannon, Touchet, Walla Walla, Umatilla, John Day, Deschutes, and Klickitat Rivers.

The Columbia Plateau is underlain by massive basalt flows having an estimated composite thickness of at least 14,000 ft at one of the plateau's lowest points near Pasco, Washington (Drost and others, 1990; Reidel and others, 2002). Sedimentary deposits overlie the basalt over large areas of the plateau, exceeding 2,100 ft in thickness in the Yakima River valley (Jones and others, 2006) and 2,000 ft in thickness in the Grande Ronde valley near La Grande, Oregon (Drost and others, 1990).

The Columbia Plateau was divided into three informal physiographic subprovinces by Myers and Price (1979)—the Yakima Fold Belt, Blue Mountains, and Palouse subprovinces; these were later modified by Reidel and others (2002), who also added an additional subprovince, the Clearwater Embayment (fig. 1). The Yakima Fold Belt includes most of the western half of the plateau north of the crest of the Blue Mountains and is characterized by a series of east-west trending anticlinal ridges and synclinal basins resulting from north-south compression following emplacement

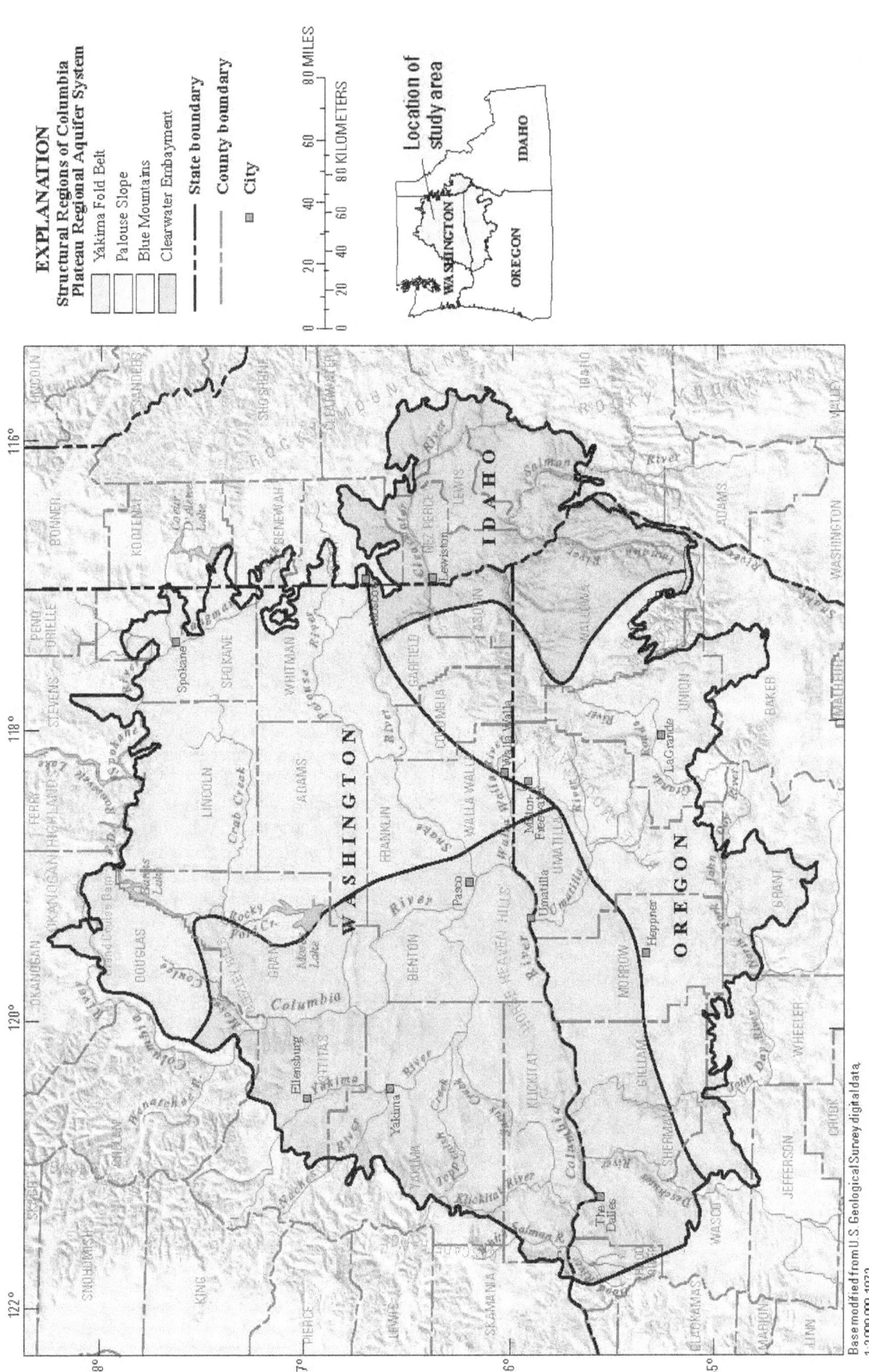

Figure 1.   Columbia Plateau Regional Aquifer System study area and structural regions (modified from Reidel and others, 2002), Washington, Oregon, and Idaho.

of the CRBG. The Palouse Slope occupies the northeast quarter of the plateau in Washington, north of the Blue Mountains, and extends eastward into Idaho. It consists of nearly undeformed basalt with a gentle southwest slope, on which a rolling topography of loess covered hills developed. The Blue Mountains subprovince comprises part of the Columbia Plateau in Oregon and Washington and includes the Blue Mountains, a composite anticlinal structure, and surrounding areas. This subprovince is characterized by higher plateaus, deeply dissected by many streams. The Clearwater Embayment marks the eastward extent of the CPRAS along the foothills of the Rocky Mountains and includes a series of folds extending into Idaho beyond Lewiston. The eastern-most extent of this subprovince extends into the Rocky Mountains along the Clearwater River drainage.

The surface topography in central Washington, commonly referred to as the "channeled scablands," was produced during Pleistocene time. Catastrophic floods, resulting from the breakup of glacial-ice dams that impounded large lakes in western Montana and northern Idaho, carved spectacular erosional features into the basalt plateau. Floodwaters stripped away overlying sediments and left behind deep canyons and coulees, rugged cliffs and buttes, large gravel bars, and giant ripple marks. Thick layers of sediment were deposited in low areas where floodwaters spread, slowed, and ponded. These slack-water deposits approach a thickness of 300 ft at several localities.

Much of the Columbia Plateau is semiarid, and the mean annual precipitation for 1895–2007 (PRISM Climate Group, 2010; calculated from annual values; fig. 2) is about 17 in. or 55,000 ft³/s (about 40 million acre-ft). Mean annual precipitation ranges from less than 7 in. in the center of the study area to more than 45 in. in the surrounding forested mountains. From the east slope of the Cascade Range, at about 2,000 ft altitude (fig. 3), to the north-westernmost extent of the study area, the mean annual precipitation is more than 60 in. Precipitation is greater in the Cascade Range to the west than at similar altitudes in the Blue Mountains to the southeast (figs. 2 and 3).

The types and amounts of natural vegetation found on the Columbia Plateau are largely dependent on precipitation quantities and land-surface altitudes. In the central part of the plateau, where the land surface ranges from 350 to 2,000 ft above sea level and the precipitation ranges from 7 to 15 in/yr, the vegetation is principally sagebrush and grasslands. At altitudes ranging from 2,000 to 3,500 ft, the vegetation is typical of semiarid climates and includes both grasslands and forest. Where altitudes generally are greater than 3,500 ft, forest lands predominate and small perennial streams in deep canyons are common. The mountainous topography is typically rugged and steep with a patchwork of barren rock and conifer forests. The generalized land cover and land use in the study area displays the regional pattern of vegetation (fig. 4). Dryland agriculture principally includes winter and spring wheat and lentils. Irrigated agriculture includes the Nation's largest production of apples and hops, as well as other crops including potatoes, onions, mint, and increasingly, wine grapes.

Geologic Setting

The geology of the Columbia Plateau, which has been studied at various levels of detail for more than 100 years, has been described in numerous reports. For the purposes of this study, a brief description of the geologic setting has been compiled from several sources relying mostly on Drost and others (1990), Reidel and others (2002), and Conlon (2006). For more detailed accounts, the reader is referred to two on-line bibliographies for lists of references that address the geology of the Columbia Plateau: Conlon (2006, http://or.water.usgs.gov/projs_dir/crbg/, accessed April 29, 2008) and specifically for the Palouse Region; Bush and others (1999, http://www.webs.uidaho.edu/pbac/pubs/biblio99.pdf, accessed March 3, 2009).

The Columbia Plateau is an intermontane basin between the Rocky Mountains and the Cascade Range filled mostly with Cenozoic basalt and sediment. The CRBG consists of a series of flows erupted during various stages of the Miocene Age, 17 million to 6 million years ago. The basalt lava flowed from fissures and vents in eastern Washington, northeastern Oregon, and western Idaho, forcing the ancient Columbia River into its present course. The number, extent, and thickness of flows vary depending on many factors, including proximity to and volume of eruption, lava viscosity, cooling process, erosion, and topography over which the lava flowed (Swanson and others, 1979; Conlon, 2006). More than 300 flows have been identified, and individual flows range in thickness from 10 to more than 300 ft (Tolan and others, 1989; Drost and others, 1990). Total thickness of the series of flows may be greater than 14,000 ft near Pasco, Washington (Reidel and others, 2002). Typically, lava erupted quickly and advanced away from the fissure or vent as a single, uniform sheet of lava. When the hiatus between flows was sufficiently long, soil developed or sediments were deposited on the surface of a flow. If these sediments were preserved, a sedimentary interbed occurs between flows.

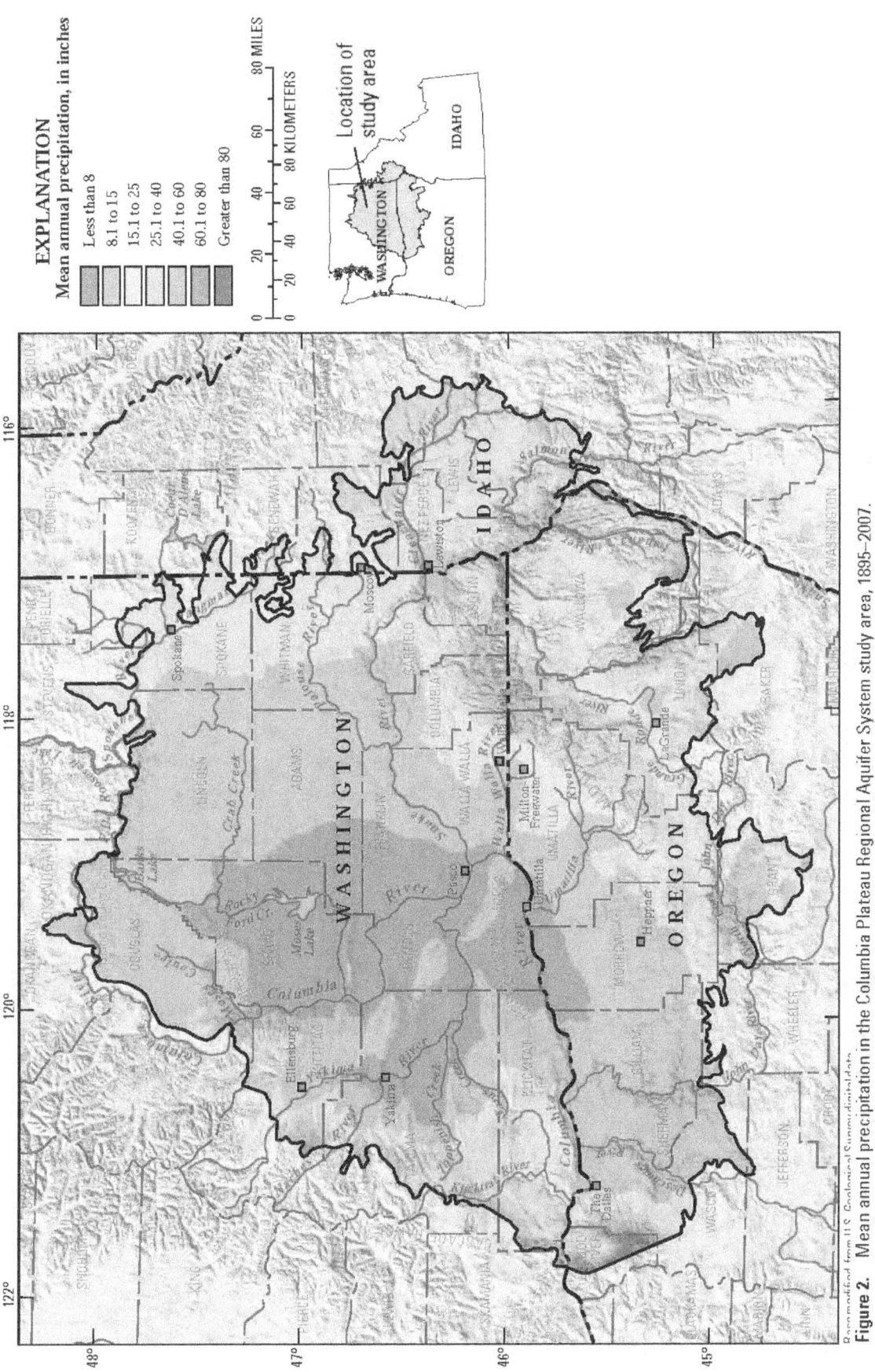

Figure 2. Mean annual precipitation in the Columbia Plateau Regional Aquifer System study area, 1895–2007.

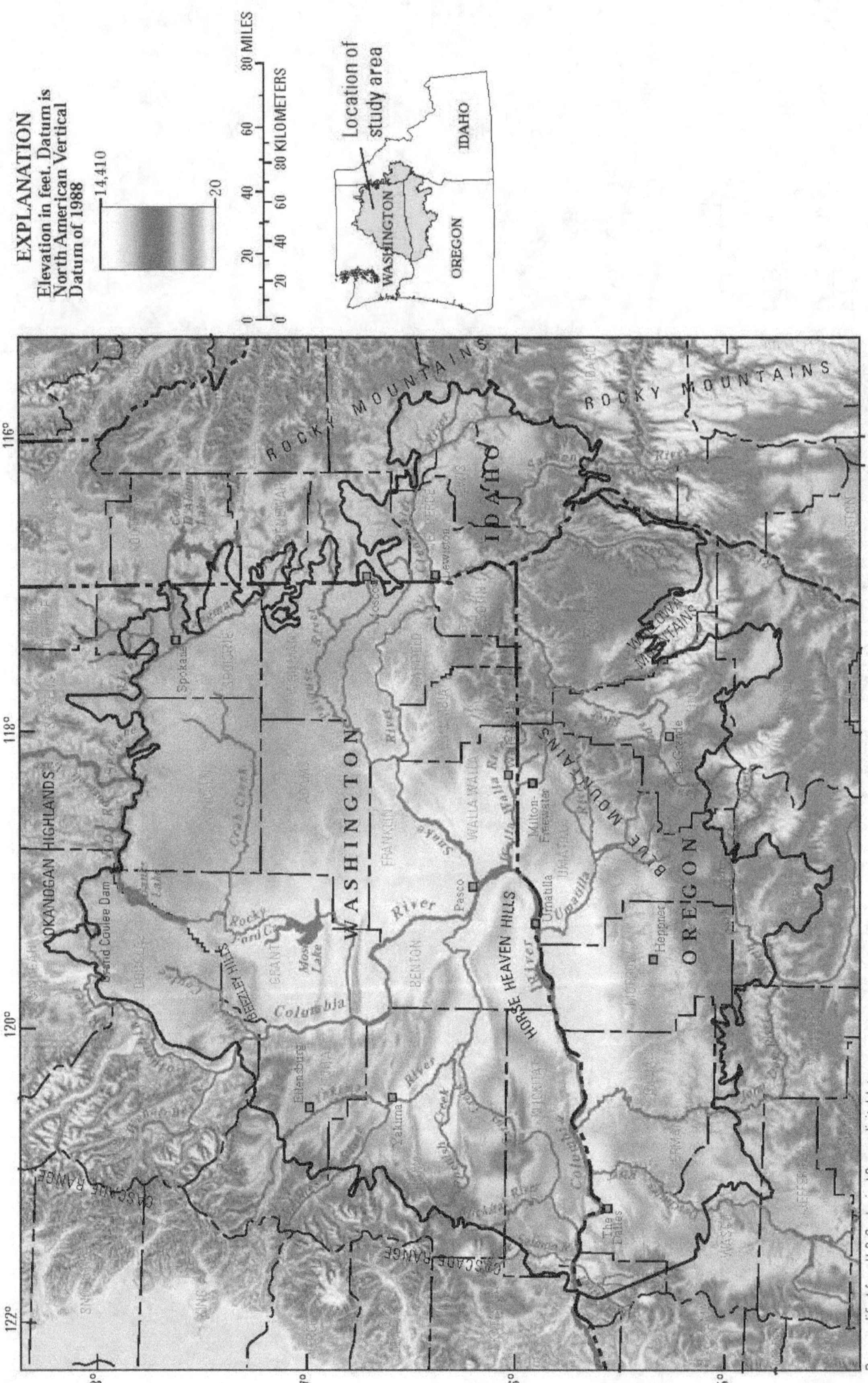

Figure 3.   Shaded relief of parts of Washington, Oregon, and Idaho.

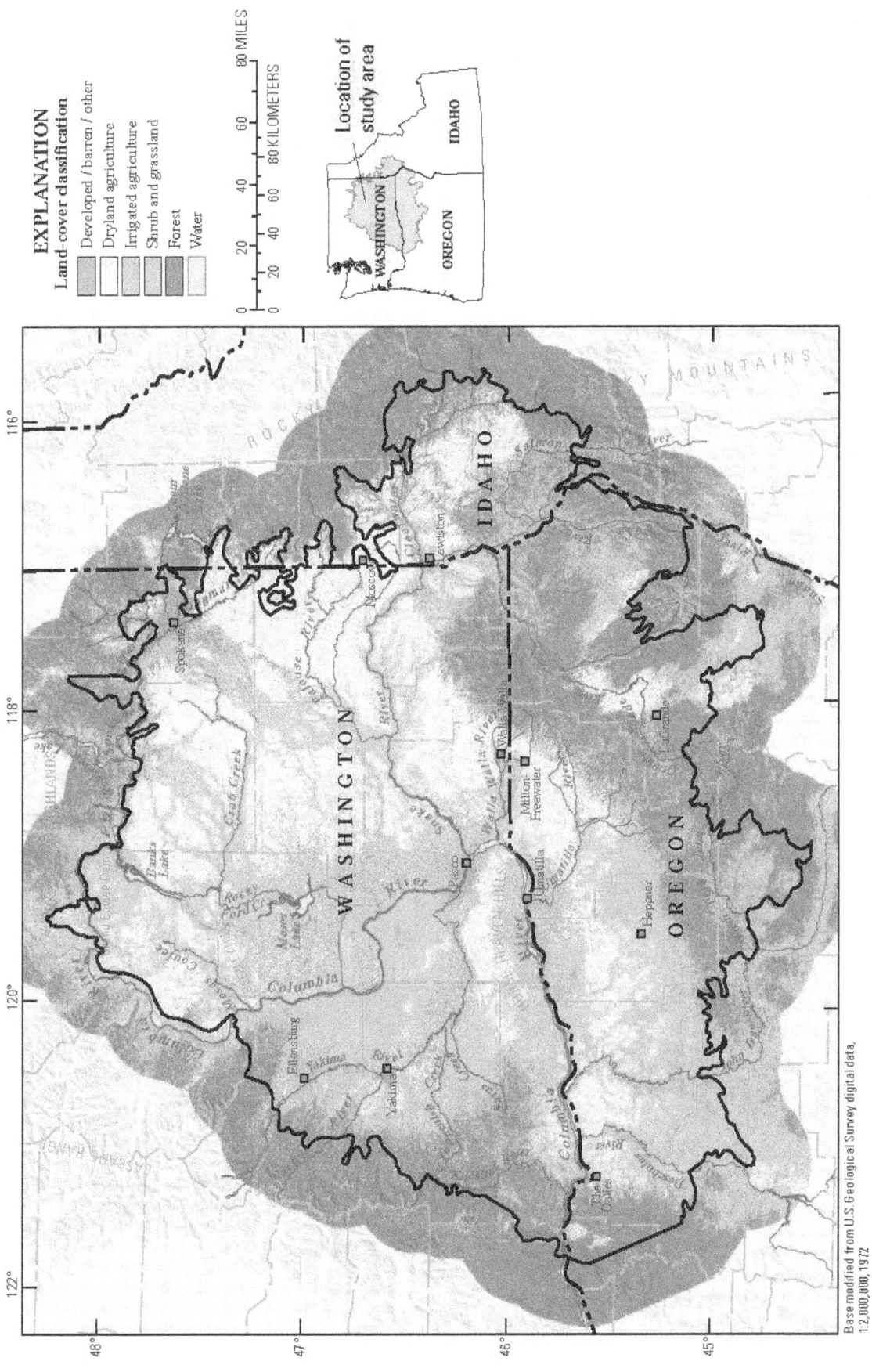

Figure 4.   Generalized land cover and land use in study area.

Base modified from U.S Geological Survey digital data,
1:2,000,000, 1972

EXPLANATION
Land-cover classification

- Developed / barren / other
- Dryland agriculture
- Irrigated agriculture
- Shrub and grassland
- Forest
- Water

Location of study area

Major CRBG-flow features are illustrated in figure 5, which typically include a permeable flow top; a dense, low permeability flow interior; and a flow bottom of variable thickness (Reidel and others, 2002). Basalt intraflow structures (vesicular- and (or) brecciated-flow tops, flow-bottom pillow complexes, and (or) brecciated zones) serve as the major aquifers in the region, while the dense-flow interiors commonly act as aquitards (Reidel and others, 2002). The zone between two individual basalt flows (excluding basalt Formation contacts) is referred to as an interflow zone and includes a flow top, the overlying basalt-flow bottom, and an intervening sedimentary interbed, if present.

Stratigraphy

The simplified stratigraphy that comprises the CPRAS is summarized in table 1. The majority of rocks exposed at land surface in the region are the CRBG, intercalated sedimentary rocks of the Ellensburg Formation, younger sedimentary rocks and deposits, Pleistocene cataclysmic flood deposits, eolian deposits, terrace gravels of modern rivers, and other localized deposits.

Sediment Stratigraphy

Within the Yakima Fold Belt (fig. 1), Miocene sedimentary deposits of the Ellensburg Formation underlie, intercalate, and overlie the CRBG and comprise most of the thickness of the unconsolidated deposits in the basinal areas (Jones and others, 2006). These continental sedimentary deposits include fluvial sands and gravels, overbank deposits, lacustrine deposits, alluvial fan deposits, sandstone, conglomerate, and interbedded volcaniclastic sediments. In eastern Washington and west-central Idaho, sediment of the Latah Formation underlies, intercalates, and overlies the CRBG (Leek, 2006). The Latah Formation consists mostly of clay, silt, and sand deposited in drainages blocked by encroaching basalt flows. Pleistocene to Holocene sediments overlying the CRBG include flood gravels and slackwater sediments, terrace gravels of modern rivers, and eolian deposits including the Palouse Formation.

Basalt Stratigraphy

The thickest, most extensive, and hydrologically most important geologic unit in the CPRAS is the CRBG (Whiteman and others, 1994). The CRBG has been divided into six geologic formations (Swanson and others, 1979): Imnaha Basalt, Picture Gorge Basalt, Prineville Basalt, Grande Ronde Basalt, Wanapum Basalt, and Saddle Mountains Basalt. These formations are divided into members and further subdivided into flow units based on field mapping, well logs, aeromagnetic surveys, geochemistry, and magnetic polarity (Conlon, 2006).

Flows belonging to the Imnaha Basalt, the oldest known in the CRBG, are found in western Idaho and eastern Washington and Oregon (fig. 6). The Picture Gorge and Prineville Basalt formations are limited to areas in central Oregon, defining the southern extent of CRBG (fig. 6). The Grande Ronde Basalt constitutes nearly 90 percent of the volume of the CRBG (Bjornstad, 2007). Flows of the Wanapum Basalt commonly overlie the Grande Ronde Basalt in most areas. Flows of the Saddle Mountains Basalt are the least extensive and youngest of the CRBG (fig. 6). During the Pleistocene, the surface expression of the basalt was modified greatly during repeated catastrophic outburst flooding, which caused erosion of vast channels and ancient waterfalls in places, as well as removal and (or) deposition of overlying sediment.

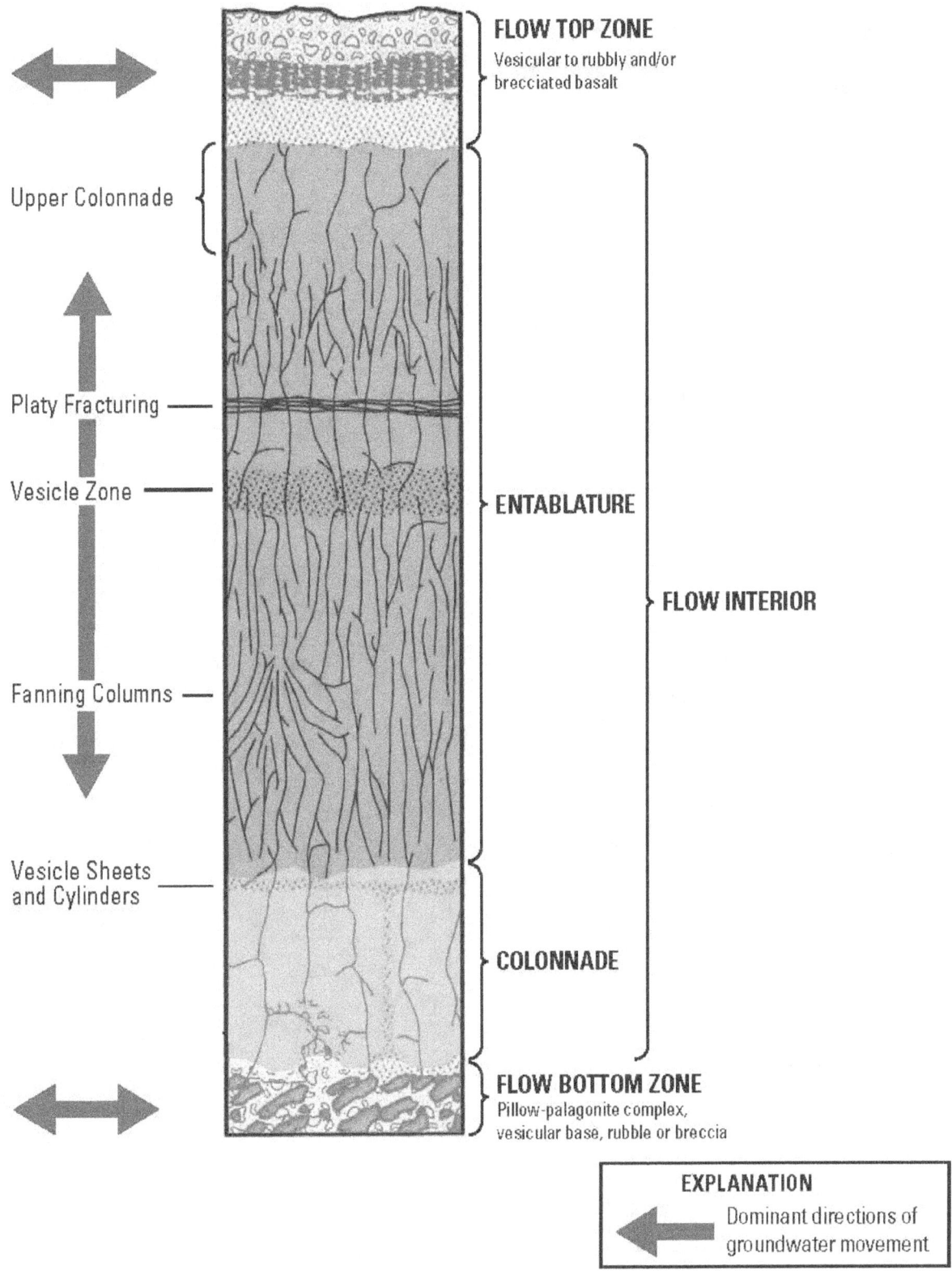

FLOW TOP ZONE
Vesicular to rubbly and/or
brecciated basalt

Upper Colonnade

Platy Fracturing

Vesicle Zone

ENTABLATURE

FLOW INTERIOR

Fanning Columns

Vesicle Sheets
and Cylinders

COLONNADE

FLOW BOTTOM ZONE
Pillow-palagonite complex,
vesicular base, rubble or breccia

EXPLANATION
Dominant directions of
groundwater movement

Figure 5. Diagram of features within a typical Columbia River Basalt Group flow (modified from Vaccaro, 1986 and Reidel and others, 2002).

Table 1. Correlation chart showing relation between generalized stratigraphy and hydrogeologic units of the Columbia Plateau Regional Aquifer System, Washington, Oregon, and Idaho (from Kahle and others, 2009).

ERA	PERIOD	EPOCH	Sediment stratigraphy	Basalt stratigraphy			Hydrogeologic unit
Cenozoic	Quaternary	Holocene	Alluvial, colluvial, eolian, glacial, glacial outburst flood, lacustrine, landslide, terrace, and peat deposits; ash, debris-avalanche and debris-flow deposits, talus; Touchet Beds, Palouse Formation	Quarternary and Pliocene Basalts			Overburden
		Pleistocene					
		Pliocene	Alluvial fan deposits; Alkali Canyon, Chenoweth, Deschutes, Madras and Ringold Formations; Dalles Group; Thorpe Gravel; and unknown continental sedimentary deposits				
	Tertiary	Miocene	Ellensburg, Deschutes, Latah, Madras, Payette, and Ringold Formations; Dalles Group; Snipes Mountain deposits; Deer Creek Beds; and unknown continental sedimentary deposits	Columbia River Basalt Group	Saddle Mountains Basalt flow members and interbeds		Saddle Mountains unit
					Mabton interbed (Mabton Member of the Ellensburg Formation)		Mabton unit
					Wanapum Basalt flow members and interbeds		Wanapum unit
					Vantage interbed (Vantage Member of the Ellensburg Formation)		Vantage unit
					Prineville Basalt	Grande Ronde Basalt flow members and interbeds	Grande Ronde unit
					Picture Gorge Basalt		
					Imnaha Basalt		
pre-Columbia River Basalt Group rocks, undivided							Older Bedrock

Names, descriptions, and ages of simplified geologic units of the Columbia Plateau.

Geologic unit symbol and color	Age	CPRAS simplified geologic map unit
Qs	Quaternary	Sediment
QTi	Quaternary-Tertiary	Intrusives
QTsdr	Quaternary-Tertiary	Sedimentary deposits or rocks
QTv	Quaternary-Tertiary	Non-CRBG volcanics
Msdr	Miocene	Sedimentary deposits or rocks
Mv(SMB)	Miocene	Saddle Mountains Basalt
Mv(WB)	Miocene	Wanapum Basalt
Mv(GRB)	Miocene	Grande Ronde Basalt
Mv(PB)	Miocene	Prineville Basalt
Mv(PGB)	Miocene	Picture Gorge Basalt
Mv(IB)	Miocene	Imnaha Basalt
Mv(CRBG)	Miocene	CRBG, undivided
preM	preMiocene	Pre-CRBG rocks, undivided

[CPRAS, Columbia Plateau Regional Aquifer System. CRBG, Columbia River Basalt Group]

EXPLANATION

Columbia Plateau Regional Aquifer System
Fault
Fold

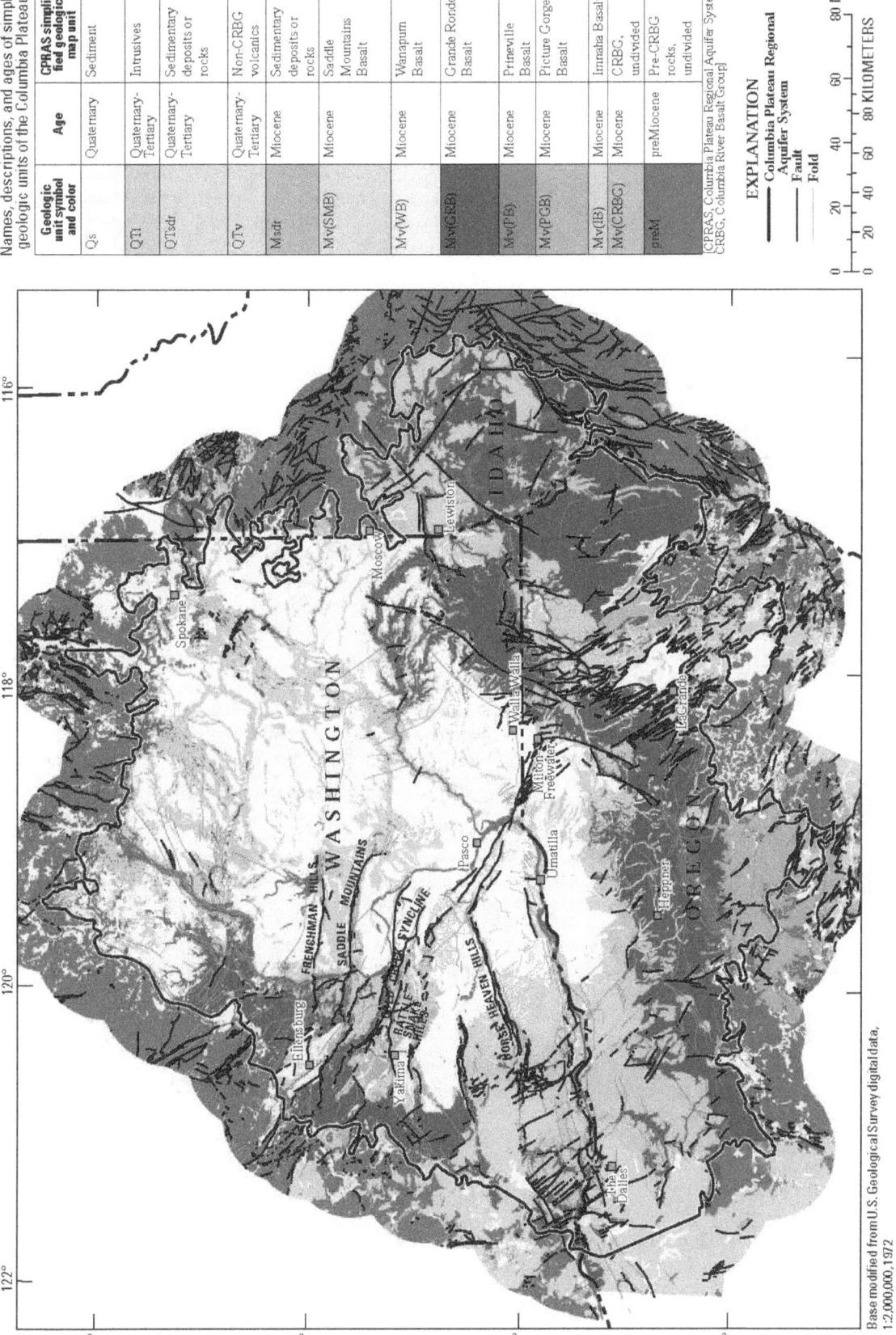

Base modified from U.S. Geological Survey digital data, 1:2,000,000, 1972

Figure 6. Simplified geologic units and major geologic structures and correlation chart showing name and age of geologic units of the Columbia Plateau Regional Aquifer System, Washington, Oregon, and Idaho (modified from Kahle and others, 2009).

Hydrogeologic Framework

Hydrogeologic Units

The conceptual groundwater model developed for the study area divides the aquifer system into seven hydrogeologic units (table 1)—the overburden aquifer, three aquifer units in the permeable basalt rock, two confining units, and the basement confining unit (Vaccaro, 1999). For the conceptual model, the three basalt hydrogeologic units are the Saddle Mountains, Wanapum, and Grande Ronde Basalts and their intercalated sediments. These basalt hydrogeologic units are distinguished from basalt formations in this study by being referred to as "units;" for example, the Wanapum Basalt and intercalated sediments are referred to as the Wanapum unit (Vaccaro, 1999). In the southeastern part of the study area, the Imnaha Basalt and any intercalated sediments are included with the Grande Ronde unit. Similarly, the Picture Gorge and Prineville Basalts in the southern part of the study area are included in the Grande Ronde unit. The confining hydrogeologic units are equivalent to the Saddle Mountains-Wanapum and Wanapum-Grande Ronde interbeds, referred to in this study as the Mabton and Vantage Interbeds, respectively (Kahle and others, 2009, table 1). The interbed units are fairly extensive laterally, but are thin when compared with the thickness of the basalt units. The basement confining unit, referred to as Older Bedrock, consists of pre-CRBG rocks that generally have much lower permeabilities than the basalts and are considered the base of the regional flow system (Vaccaro, 1999; Kahle and others, 2009). A summary description of the hydrogeologic units in the study area is provided below. The approximate surficial distribution of overburden and the three basalt hydrogeologic units are shown in figure 7A. The approximate subsurface distribution of the CPRAS hydrogeologic units is shown in figure 7B. Detailed descriptions of the units are available in Drost and others (1990), Whiteman and others (1994), Vaccaro (1999), and Jones and Vaccaro (2008).

Overburden Unit

The Overburden unit consists of undivided, unconsolidated to semi-consolidated sedimentary deposits and minor basalt and andesite, ranging from Miocene to Holocene in age (Drost and others, 1990). In it are grouped several formations of local and (or) regional extent and numerous types of deposits including alluvial, colluvial, eolian, glacial, glacial outburst flood, lacustrine, landslide, terrace, and peat deposits; talus; and other unknown continental sedimentary deposits (table 1). Thickness of the Overburden unit ranges from 0 to 1,300 ft, with a median thickness of about 47 ft (Kahle and others, 2009).

Saddle Mountains Unit

The Saddle Mountains unit consists mostly of the Saddle Mountains Basalt and interbed members. Most of the unit is in the west-central part of the study area, with less continuous occurrences in the Blue Mountains and eastward into Idaho (fig. 6). Kahle and others (2009) estimated an areal extent of about 8,000 mi², with a range in altitude of the unit top from about 4,000 to -280 ft above sea level. Thickness of the Saddle Mountains unit ranges from about 0 to 990 ft, with a median thickness of 280 ft (Kahle and others, 2009).

Mabton Interbed Unit

The Mabton unit is the sedimentary interbed between the overlying Saddle Mountains unit and the underlying Wanapum unit. The Mabton unit consists of the Mabton Member of the Ellensburg Formation and is mostly in the west-central part of the study area. Limited surficial outcrops of the Mabton unit are present in the study area and the extent of the Mabton unit is assumed to be within the extent of the Saddle Mountains unit. The Mabton unit generally consists of clay, shale, claystone, clay with basalt, clay with sand, and sandstone. Thickness of the Mabton unit ranges from about 0 to 520 ft, with a median thickness of about 44 ft (Kahle and others, 2009).

Wanapum Unit

The Wanapum unit, composed mostly of basalt and interbed members of the Wanapum basalt, is in most of the north-central part of the study area (fig. 6) and has an estimated areal extent of about 25,000 mi² with the elevation of the top ranging from about 3,400 to -1,000 ft relative to sea level (Kahle and others, 2009). Much of the unit lies beneath the Overburden and Saddle Mountains units. Thickness of the Wanapum unit ranges from about 0 to 1,200 ft, with a median thickness of about 330 ft (Kahle and others, 2009).

Vantage Interbed Unit

The Vantage unit is the sedimentary interbed between the overlying Wanapum unit and the underlying Grande Ronde unit. Over most of the study area, this unit consists of the Vantage Member of the Ellensburg Formation; however, this unit includes sediment of the Latah Formation in the northeastern part of the study area. Limited surficial outcrops of this unit are present in the study area and the extent is assumed to be within the extent of the Wanapum unit.

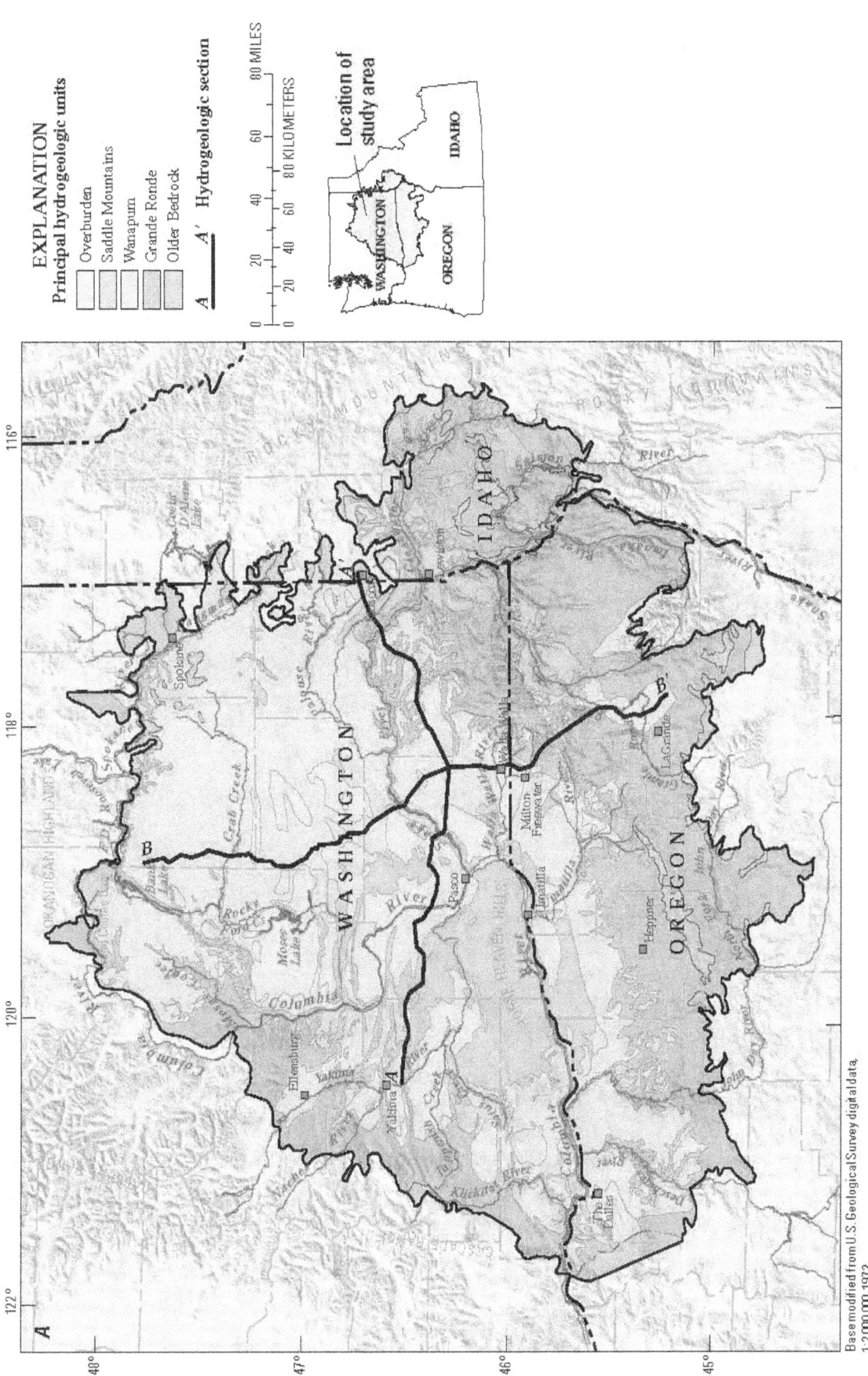

Figure 7A. Surficial distribution of overburden and basalt hydrogeologic units in the Columbia Plateau Regional Aquifer System, Washington, Oregon, and Idaho (from Kahle and others, 2009).

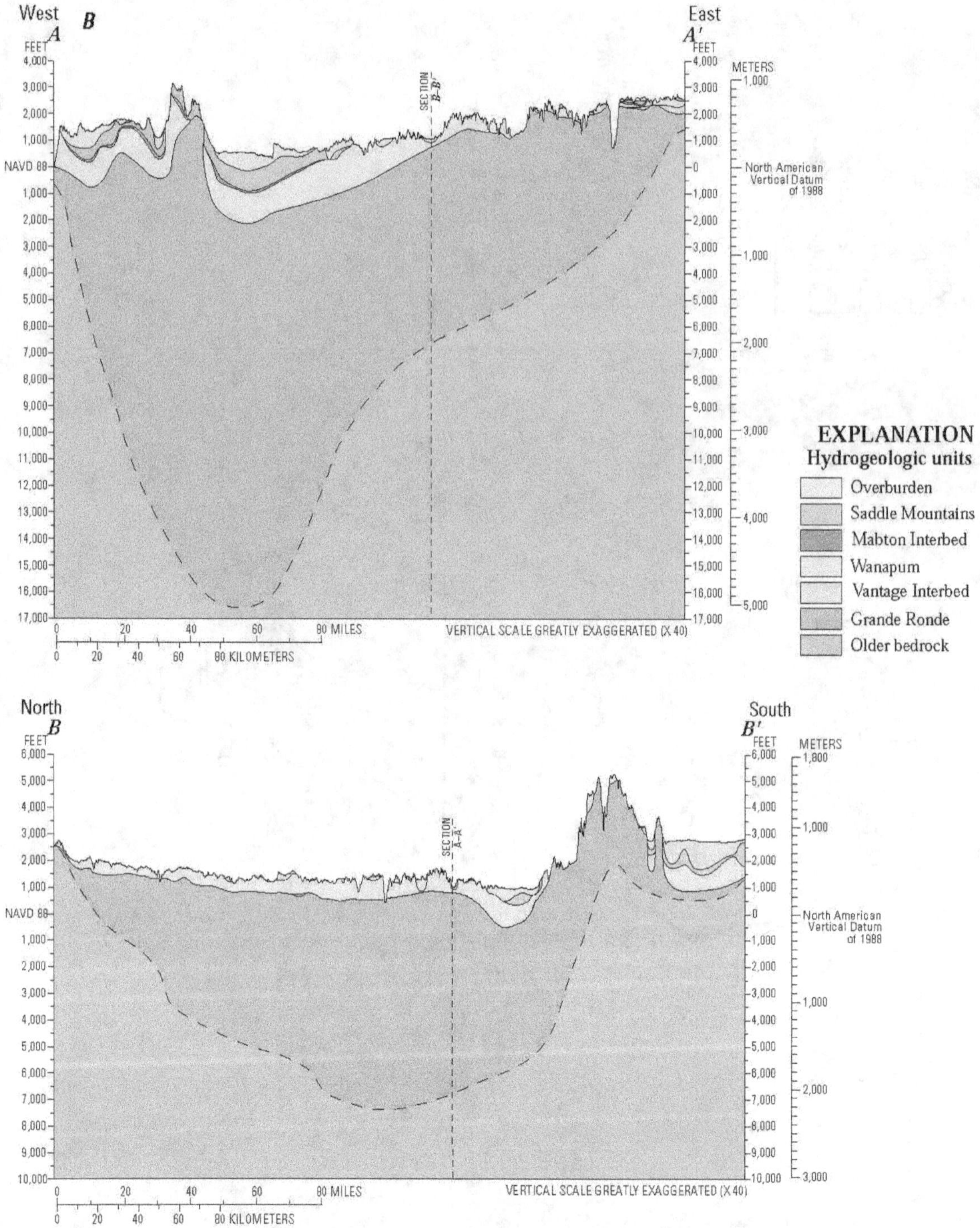

EXPLANATION
Hydrogeologic units

- Overburden
- Saddle Mountains
- Mabton Interbed
- Wanapum
- Vantage Interbed
- Grande Ronde
- Older bedrock

Figure 7B.   Generalized hydrogeologic sections of the Columbia Plateau Regional Aquifer System, Washington, Oregon, and Idaho (from Kahle and others, 2009).

The Vantage unit consists of clay, shale, sandstone, tuff with claystone, and clay with basalt, but also may contain small amounts of sand and sand-and-gravel. A few well logs also indicate that the Vantage unit is not present in the southeastern part of the Yakima River Basin and near the Cold Creek Syncline and Rattlesnake Hills Structure (Jones and Vaccaro, 2008). Thickness of the Vantage unit ranges from about 0 to 320 ft, with a median thickness of about 20 ft (Kahle and others, 2009).

Grande Ronde Unit

The Grande Ronde unit is the oldest and most extensive of the basalt units. This unit underlies most of the study area, except for an area along the southern boundary of the CPRAS in Oregon and along the eastern extent in Idaho (fig. 6). The estimated areal extent of the Grande Ronde unit is about 42,000 mi² (Kahle and others, 2009). The Grande Ronde unit predominantly contains the basalt and interbed members associated with the Grande Ronde Basalt. Sedimentary interbeds within the unit generally are rare and where present are only a few feet thick. The top of the Grande Ronde unit ranges from 4,300 to -2,100 ft based on the well log data used by Kahle and others (2009). Thickness of the unit is largely unknown, but is estimated to be greater than 14,000 ft near the central part of the basin.

Older Bedrock Unit

The Older Bedrock unit that borders and underlies the CPRAS is composed of various rock types older than the CRBG (Kahle and others, 2009). In Washington and Idaho, the rocks bordering the CPRAS consist mostly of sedimentary and granitic rocks. In Oregon, the CPRAS is bordered by sedimentary, volcaniclastic, volcanic, plutonic, and metamorphic rocks (Drost and others, 1990).

Three-Dimensional Hydrogeologic Model

As part of the CPRAS groundwater-availability study, a three-dimensional geologic model (Burns and others, 2011) was constructed in order to define the general aquifer system geometry for use in the regional numerical groundwater-flow model that is being developed as the end product of this study. The geologic-model units consist of the CRBG and overlying sediments described above using data compiled from numerous databases and detailed studies that were completed during the past three decades. These data include

stratigraphic interpretations of more than 13,000 wells and a contiguous compilation of surficial geology and structural features in the study area. These data were simplified and used to construct piecewise-smooth trend surfaces that represent upper and lower subsurface-model unit boundaries. These smooth surfaces were generated using LOESS (Cleveland and others, 1992) trend modeling methods to decompose the data into well-supported trends and apparently random residuals (Burns and others, 2011). Surfaces were recombined using a rule-based algorithm to construct a fully three-dimensional model with a 500-ft grid resolution that is consistent with the data. The modeling process yielded improved estimates of unit volumes, refinement of location of large structural features, and identification of features that may be important for ongoing groundwater studies. An on-line interactive tool was developed to serve point information and cross sections developed from the geologic model to the general public (http://or.water.usgs.gov/proj/cpras/index.html, accessed March 23, 2011).

Hydraulic Characteristics of Units

The ability of sediments and rocks to store and transmit groundwater (their hydraulic characteristics) determines how a groundwater-flow system functions. Knowledge of the hydraulic characteristics also is necessary to evaluate how the flow system responds to stresses such as pumpage. These characteristics include lateral and vertical hydraulic conductivity and the storage coefficient. Estimates of characteristics from previous studies and this study are described below. This information provides a general overview of the range and median of hydraulic-characteristic values for the hydrogeologic units in the CPRAS.

Most of the information presented in this section was originally compiled, analyzed, and reported on during a study of the hydrogeologic framework of the Yakima River Basin (Vaccaro and others, 2009), an effort that included compilation of hydraulic characteristics throughout the entire CPRAS.

Lateral Hydraulic Conductivity

Lateral hydraulic conductivity (referred to herein as K_h) is a measure of a material's ability to transmit water laterally. It is expressed in units of cubic feet per square feet per day—simplified to feet per day. Values of K_h can be estimated from specific-capacity data reported on drillers' logs or determined from aquifer tests or groundwater-flow modeling. Numerous studies conducted within the CPRAS have calculated or compiled information on K_h (table 2).

Table 2.   Summary of previous estimates of lateral hydraulic conductivity, Columbia Plateau Regional Aquifer System, Washington, Oregon, and Idaho.

[Abbreviations: ft, foot; <, less than; –, not applicable]

Overburden unit	Basin or area	Minimum	Median	Mean	Maximum	Model derived	Source
Alluvium	Toppenish	95	–	–	780	–	Bolke and Skrivan, 1981
	Yakima	–	–	–	–	280	Golder Associates, 2002
Alluvium, fluvial gravels, coarse sands	Pasco/Benton	10	–	280	1,000	–	Zimmerman, 1983
Alluvium, coarse to medium grained (some fines and caliche)	Hanford	6	–	–	100,000	–	Vermeul and others, 2001
	Hanford/Black Rock	–	–	0.85	–	–	Didricksen, 2004
Alluvium, unconfined	Ellensburg	–	–	2,000	–	–	GeoEngineers, 1999
Alluvium, unconsolidated	Yakima	670	–	2,000	802	–	Foxworthy, 1962
Alluvium, young	Lower Satus Creek	–	–	–	–	860	Prych, 1983
Alluvium/Ellensburg (upper)	Yakima	–	–	–	–	75	Golder Associates, 2002
Alluvium, old	Lower Satus Creek	–	–	–	–	86	Prych, 1983
Basin-fill units	Yakima	0.01	6	182	18,000	–	Vaccaro and others, 2009
Overburden aquifer	Columbia Plateau	0.04	43	99	1,100	–	Hansen and others, 1994
	Columbia Plateau	0.02	240	–	150,000	–	Whiteman and others, 1994
Wind-blown deposits (loess)	Columbia Plateau	1	–	–	–	5	Lum and others, 1990
	Columbia Plateau	–	–	–	10	–	Vaccaro, 1999
Plio-Pleistocene Unit	Hanford	0.02	–	–	57	–	Khaleel and Freeman, 1995
Sediments	Horse Heaven Hills	–	–	24,000	–	–	Davies-Smith and others, 1989
Ellensburg Formation, upper (old Alluvium)	Lower Satus Creek	0.9	72	–	3,000	–	Prych, 1983
Ellensburg Formation, upper	Ellensburg	2	–	–	–	53	GeoEngineers, 1999
Ellensburg Formation, upper (all)	Yakima	0.6	–	13	2,300	–	Golder Associates, 2002
Ellensburg Formation, upper (upper)	Yakima	–	–	–	–	7.5	Golder Associates, 2002
Ellensburg Formation, upper (upper 300 ft)	Ellensburg	–	–	–	–	32	GeoEngineers, 1999
Ellensburg Formation, upper (upper and middle)	Yakima	0.6	–	12	2,300	–	Golder Associates, 2002
Ellensburg Formation, upper (middle)	Yakima	–	–	–	–	0.002	Golder Associates, 2002
Ellensburg Formation, upper (lower)	Yakima	38	–	60	96	–	Golder Associates, 2002
	Yakima	–	–	–	–	8	Golder Associates, 2002
	Yakima	–	–	–	–	0.1	Golder Associates, 2002
	Yakima	–	–	–	–	0.01	Golder Associates, 2002
Slack water deposits (Touchet beds)	Pasco/Benton	8	–	–	11	9	Drost and others, 1997
Soil buried horizon, caliche, gravel	Hanford	47	–	–	620	–	Vermeul and others, 2001
Soils, Palouse	Hanford	0.2	–	–	0.4	–	Khaleel and Freeman, 1995
Glaciofluvial deposits (Pasco gravel)	Pasco/Benton	48	880	–	73,000	880	Drost and others, 1997
	Pasco/Benton	–	1,200	–	–	–	Meyers and others, 1985
Hanford Formation	Hanford	3	–	–	3,300,000	–	Wurstner and others (1995); Thorne and Newcomer (1992)

Table 2. Summary of previous estimates of lateral hydraulic conductivity, Columbia Plateau Regional Aquifer System, Washington, Oregon, and Idaho.—Continued

[Abbreviations: ft, foot; <, less than; –, not applicable]

Basin or area	Lateral hydraulic conductivity, in feet per day					Source
	Minimum	Median	Mean	Maximum	Model derived	
Overburden unit—Continued						
Ringold Formation — Hanford/Black Rock	–	2.6	–	–	–	Didricksen, 2004
Ringold Formation, upper — Pasco/Benton	2	25	–	210	5	Drost and others, 1997
Hanford	–	–	–	–	0.02	Vermeul and others, 2001
Hanford	0.001	–	–	0.3	–	Wurstner and others (1995); Thorne and Newcomer (1992)
Ringold Formation, middle — Pasco/Benton	200	–	–	250	–	Bergeron and others, 1986
Pasco/Benton	8	180	–	5,000	200	Drost and others, 1997
Pasco/Benton	–	41	–	–	–	Newcomb and others (1972); Graham (1981); Meyers and others (1985)
Hanford	0.1	–	–	5,500	–	Vermeul and others, 2001
Hanford	0.03	–	–	280	–	Vermeul and others, 2001
Hanford	0.03	–	–	0.3	–	Vermeul and others, 2001
Hanford	0.3	–	–	656	–	Wurstner and others (1995); Thorne and Newcomer (1992)
Hanford	0.001	–	–	0.3	–	Wurstner and others (1995); Thorne and Newcomer (1992)
Hanford	0.3	–	–	660	–	Wurstner and others (1995); Thorne and Newcomer (1992)
Ringold Formation, middle, gravel facies — Hanford	2,000	–	–	6,400	–	Vermeul and others, 2003
Ringold Formation, middle, sand facies — Hanford	–	–	1,300	–	–	Vermeul and others, 2003
Ringold Formation, middle, silt facies — Hanford	–	–	2	–	–	Vermeul and others, 2003
Ringold Formation, lower — Pasco/Benton	2	46	–	230	4	Drost and others, 1997
Hanford	0.001	–	–	0.3	–	Wurstner and others (1995); Thorne and Newcomer (1992)
Ringold, lower, basal, clay/mud — Hanford	–	–	–	–	3.0E-5	Vermeul and others, 2001
Ringold Formation, basal — Pasco/Benton	200	–	–	250	–	Bergeron and others, 1986
Hanford	0.3	–	–	660	–	Wurstner and others (1995); Thorne and Newcomer (1992)
Ringold, basal, sand/gravel — Hanford	0.03	–	–	700	–	Vermeul and others, 2001
Basalt units						
Rattlesnake Ridge Interbed (within upper Saddle Mountains) — Hanford/Black Rock	–	–	0.8	–	–	Didricksen, 2004
Hanford	0.05	–	8.4	210	–	Strait and Mercer, 1987
Selah Interbed (within upper Saddle Mountains) — Hanford/Black Rock	–	–	2.7	–	–	Didricksen, 2004

Table 2. Summary of previous estimates of lateral hydraulic conductivity, Columbia Plateau Regional Aquifer System, Washington, Oregon, and Idaho.—Continued

[Abbreviations: ft, foot; <, less than; –, not applicable]

	Basin or area	Lateral hydraulic conductivity, in feet per day					Source
		Minimum	Median	Mean	Maximum	Model derived	
Basalt units—Continued							
Saddle Mountains	Horse Heaven Hills	–	–	18	–	–	Davies-Smith and others, 1988
	Umatilla/Horse Heaven Hills	–	–	–	–	8.0	Davies-Smith and others, 1988
	Pasco/Benton	0.007	2	–	3,200	–	Drost and others, 1997
	Columbia Plateau	0.2	1	1	3	–	Hansen and others, 1994
	Columbia Plateau	0.007	2	–	1,900	–	Whiteman and others, 1994
	Pasco/Benton	–	–	–	–	1	Zimmerman, 1983
	Hanford/Black Rock	0.04	–	–	2.6	–	Didricksen, 2004
	Hanford	0.04	2	10	53	–	Schmidt and others, 2007
Saddle Mountains anticline	Horse Heaven Hills	0.03	–	–	0.5	–	Packard and others, 1996
	Horse Heaven Hills	0.2	–	–	78	–	Packard and others, 1996
Mabton Interbed	Hanford/Black Rock	–	–	0.025	–	–	Didricksen, 2004
Wanapum	Horse Heaven Hills	–	–	170	–	–	Davies-Smith and others, 1988
	Umatilla/Horse Heaven Hills	–	–	–	–	4.7	Davies-Smith and others, 1988
	Pasco/Benton	1.1	11	–	310	–	Drost and others, 1997
	Columbia Plateau	0.9	–	–	1	–	Lum and others, 1990
	Columbia Plateau	0.09	3	3	8	–	Hansen and others, 1994
	Horse Heaven Hills	0.1	–	–	260	–	Packard and others, 1996
	Columbia Plateau	0.007	5	–	5,200	–	Whiteman and others, 1994
	Palouse	0.01	–	–	1,000	–	Golder and Associates, 2004
	Pasco/Benton	–	–	–	–	5	Zimmerman, 1983
Wanapum, interflow	Hanford	–	2.8	–	–	–	Reidel and others, 2002
Wanapum, flow interior	Hanford	–	3.0E-6	–	–	–	Spane and Thorne, 1985
Grande Ronde	Horse Heaven Hills	–	–	65	–	–	Davies-Smith and others, 1988
	Umatilla/Horse Heaven Hills	–	–	–	–	1.3	Davies-Smith and others, 1988
	Columbia Plateau	1	–	–	12	–	Lum and others, 1990
	Columbia Plateau	0.005	5	–	5,200	–	Whiteman and others, 1994
	Palouse	1.0E-4	–	–	0.1	–	Golder Associates, 2004
	Pasco/Benton	–	–	–	–	0.02	Zimmerman, 1983
Grande Ronde, top	Hanford	3.0E-6	0.1	–	500	–	Strait and Mercer, 1987
Grande Ronde, 1–2,000 ft thick	Columbia Plateau	0.1	1	2	9	–	Hansen and others, 1994
Grande Ronde, flow interior	Hanford	5.0E-9	2.0E-7	–	0.001	–	Strait and Mercer, 1987
Grande Ronde, flow interior	Hanford	1.0E-8	1.4E-7	–	1.7E-5	–	Eslinger, 1986
Grande Ronde, interflow	Hanford	–	0.03	–	–	–	Reidel and others, 2002

Table 2. Summary of previous estimates of lateral hydraulic conductivity, Columbia Plateau Regional Aquifer System, Washington, Oregon, and Idaho.—Continued

[Abbreviations: ft, foot; <, less than; —, not applicable]

Basin or area	Minimum	Median	Mean	Maximum	Model derived	Source
Basalt units—Continued						
Grande Ronde, greater than 2,000 ft thick — Columbia Plateau	0.1	1	2	6	—	Hansen and others, 1994
Basalt — Yakima	—	—	—	—	5	Golder Associates, 2002
Basalt — Columbia Plateau	0.9	—	—	9	—	Bolke and Vaccaro, 1981
Basalt — Columbia Plateau	0.005	5	—	6,100	—	Whiteman and others, 1994
Basalt, top — Pasco/Benton	—	35	—	—	—	Drost and others, 1997
Basalt, center — Pasco/Benton	—	2.0E-7	—	—	—	Drost and others, 1997
Basalt, interflow — Hanford	3.0E-7	—	—	2,800	—	Reidel and others, 2002
Basalt, flow interior — Hanford	3.0E-10	3.0E-7	—	0.003	—	Reidel and others, 2002
Basalt, weathered — Yakima	—	—	—	—	0.003	Golder Associates, 2002
Basalt, undisturbed — Pasco/Benton	0.03	—	—	8	—	Zimmerman, 1983
Basalt, anticline — Pasco/Benton	0.004	—	—	0.3	—	Zimmerman, 1983
Basalt, fault — Pasco/Benton	1.0E-4	—	—	0.01	—	Zimmerman, 1983
Basalt, barrier (flow impediment) — Pasco/Benton	0.002	—	—	0.05	—	Zimmerman, 1983
Columbia River Basalt Group units — Yakima	0.1	3	180	49,000	—	Vaccaro and others, 2009
Columbia River Basalt — Willamette	22	—	—	1,100	—	Conlon and others, 2005
Columbia River Basalt — Willamette	—	6	—	—	—	Woodward and others, 1998
Columbia River Basalt — Willamette	—	—	—	—	2.5	Woodward and others, 1998
Columbia River Basalt — Portland	—	—	—	—	0.1-3	Morgan and McFarland, 1996
Columbia River Basalt — Portland	<0.001	0.3	—	200	—	McFarland and Morgan, 1996
Columbia River Basalt, structurally affected — Willamette	0.001	—	—	0.1	—	Woodward and others, 1998
Columbia River Basalt, undeformed — Willamette	1	—	—	3	—	Woodward and others, 1998
Columbia River Basalt — Columbia Plateau	0.001	1	—	750	—	Woodward and others, 1998
Pomona and Priest Rapids — Mosier	—	500-1,000	—	—	—	Lite and Grondin, 1988
Pomona flow top — Mosier	—	—	—	—	500	Burns, 2011, U.S. Geological Survey, written commun.
Pomona flow bottom — Mosier	—	—	—	—	50	Burns, 2011, U.S. Geological Survey, written commun.
Lolo flow top — Mosier	—	—	—	—	100	Burns, 2011, U.S. Geological Survey, written commun.
Rosalia flow top — Mosier	—	—	—	—	10	Burns, 2011, U.S. Geological Survey, written commun.
Lumped Frenchman Springs flow tops — Mosier	—	—	—	—	1	Burns, 2011, U.S. Geological Survey, written commun.
Bedrock units						
Bedrock units — Yakima	0.01	3	13	40	—	Vaccaro and others, 2009
Bedrock, crystalline — General	3.0E-6	—	—	0.3	—	Trainer, 1988
Bedrock, unfractured metamorphic and igneous rocks — General	3.0E-8	—	—	3.0E-5	—	Freeze and Cherry, 1979

Lateral hydraulic conductivity, in feet per day

Overburden Unit

Overburden deposits are diverse in lithology and, thus, so are their hydraulic characteristics (table 2). The deposits, which consist of unconsolidated and consolidated material of alluvial, glacial, lacustrine, wind-blown, and volcanic origins, form important water-bearing units, as well as semiconfining to confining units. Estimates of K_h for the deposits within the Overburden unit ranged from 0.001 to 3,300,000 ft/d. This large range in estimated values is owing to the large variation in grain size, depositional regimes, and age of the deposits. The median reported values were in the range of 6 to 1,200 ft/d (table 2).

K_h of loess, which mantels the eastern part of the study area, is between 1 to 10 ft/d. The Touchet beds, slack-water deposits that mantel part of the study area upgradient from Wallula Gap, have estimated K_h values ranging from 8 to 11 ft/d. Values of K_h for the alluvium range from 6 to 100,000 ft/d and generally average from 300 to 2,000 ft/d. The large range is owing to the variation in grain size (silty sands to cobbles) of the alluvium. Estimates of K_h for the Pasco gravels, glaciofluvial sediment near Pasco, ranges from about 48 to 73,000 ft/d, with a median reported value from about 880 to 1,200 ft/d. The reported values for the upper, middle, lower, and basal Ringold Formation ranges from about 0.001 to 6,400 ft/d, reflecting a wide range in local conditions.

K_h of the Ellensburg Formation ranges from about 0.6 to 3,000 ft/d. The large range in values is owing to the variations in the types of materials composing the water-producing zones in the Ellensburg Formation. These materials range from sandstone to un-cemented sands and gravels.

Columbia River Basalt Group Units

Hydraulic characteristics vary greatly within and between the individual basalt flows, members, and hydrogeologic units (table 2). Upper zones of the flows were exposed to weathering processes and were broken by subsequent flows, resulting in the formation of conductive "flow tops." In general, the flow tops are brecciated and (or) vesicular, and the flow bases are brecciated and may contain pillow complexes if the basalt was extruded within or flowed into water. These flow tops, when combined with the base of the overlying basalt flow, form interflow zones that generally exhibit high K_h (Lindolm and Vaccaro, 1988). The interflow zones are separated by the less-transmissive flow interior (fig. 5) in which the fractures typically are vertically oriented (Tomkeieff, 1940; Waters, 1960; MacDonald, 1967; Swanson and Wright, 1978; Sublette, 1986; Hansen and others, 1994;

Whiteman and others, 1994). The flow interior fractures are a result of differential contraction during cooling of basalt flows (MacDonald, 1967; Long and Wood, 1986) and of later folding and faulting. The greatest density and lowest K_h generally occur in the interior or middle of a basalt flow, typically the entablature (fig. 6; Wood and Fernandez, 1988; Reidel and others, 2002). Observations of exposed colonnades, which typically are three- to eight-sided columns of basalt, indicate that there would be lateral connectivity along the columns; springs have been observed emanating from exposed colonnades. Many CRBG flows have been affected to some degree by fracture filling with mineral precipitates (such as smectite and clinoptilolite) that decreases K_h (Wood and Fernandez, 1988; Steinkampf and Hearn, 1996). Fractures tend to be filled with these alteration products, whereas vesicles typically are only partly filled (Steinkampf and Hearn, 1996). Such alteration products are well documented (Ames, 1980; Benson and Teague, 1982; Hearn and others, 1985; Steinkampf and Hearn, 1996).

The Saddle Mountains unit K_h has been estimated to range from 0.007 to 3,200 ft/d, with a median of about 1 to 2 ft/d. The Wanapum unit had a slightly larger range (0.007 to 5,200 ft/d) than the Saddle Mountains unit, and the median reported K_h for the Wanapum unit ranges from about 3 to 11 ft/d (excluding value for flow interior). The range in Grande Ronde unit K_h values was similar to that for the Wanapum unit, from 0.005 to 5,200 ft/d. Median K_h values for the Grande Ronde unit range from about 0.1 to 5 ft/d (again, excluding value for flow interior). Previous work indicates the CRBG K_h is one to two orders of magnitude smaller along anticlines owing to fault structures and compression of the basalts, and in deeper parts of the Grande Ronde unit (Hansen and others, 1994; Packard and others, 1996; Reidel and others, 2002) owing to overburden pressure and secondary mineralization, than in other parts of the units. For anticlines, Hansen and others (1994) reduced K_h values by multiplying the values by factors ranging from 0.01 to 0.018. For all types of flow barriers, the median multiplication factor was 0.18 (Hansen and others, 1994).

Strait and Mercer (1987) found K_h in basalt flow tops to be as much as five orders of magnitude greater than in basalt flow centers. Poeter (1980) reported that representative effective porosities ranged from about 0.02 in the flow interiors to about 0.14 in the interflow zones, further indicating the difference between K_h in interflow zones and flow interiors. Hydraulic testing of interbeds in the basalts has been limited, but reported mean values range from 0.05 to 210 ft/d.

Older Bedrock Unit

The areas bordering and underlying the CPRAS include metamorphic (crystalline), sedimentary, volcanic, and intrusive and extrusive igneous rocks. In general, the older bedrock has lower values of porosity and permeability than the overburden and CRBG units. Water-producing zones are variable, but are present in the bedrock.

Typical values of K_h for unfractured metamorphic and igneous rocks range from 3×10^{-8} to 3×10^{-5} ft/d (Freeze and Cherry, 1979). K_h values for fractured metamorphic and igneous rocks can be five orders of magnitude larger than for unfractured rock (about 0.001 to –1 ft/d). Joints within crystalline rock are of limited lateral extent but are numerous enough to increase permeability locally, and such fractures are commonly tighter and less abundant with increasing depth owing to the state of stress in the earth's crust (Trainer, 1988). The latter conditions may be important in the study area because of the existing tectonic stress that would tend to decrease fracture openings. Previous studies of well yields in crystalline rocks indicate networks of open joints are found principally within 300 to 500 ft of the surface and decline lognormally with increasing depth (Trainer, 1988). Trainer also reported that K_h values for crystalline rocks range from 3×10^{-6} to 0.3 ft/d. Thus, K_h of the Older Bedrock unit likely is quite small; many well yields are reported on driller's logs as less than 1 gal/min or 'no water.'

Estimates from Specific-Capacity Data

Lateral hydraulic conductivities were estimated as part of this study using available data for water-level change (drawdown) and discharge rate (well yield) for wells pumped for periods that ranged from 1 to 200 hours. Data from wells that had a driller's log containing a discharge rate, duration of pumping, drawdown, static water level, well-construction data, and lithologic log were used. The methods and assumptions for calculating K_h are described in appendix A. Assumptions for calculating K_h from specific-capacity data generally are not strictly met resulting in values that may be considered rough estimates. Spatial variations can provide indications of patterns of K_h, however, and the availability of many values allows for a reasonable estimate of a median value. The limitations of using specific-capacity tests are described by Gannett and Lite (2004) and Meier and others (1999).

Statistical summaries of K_h values calculated from the specific-capacity data, by hydrogeologic unit, are shown in table 3; their distribution is shown in figure 8. Mean

Table 3. Summary of lateral hydraulic conductivity values estimated from specific-capacity data, Columbia Plateau Regional Aquifer System, Washington, Oregon, and Idaho.

Hydrogeologic unit(s)	Number of wells	Hydraulic conductivity (feet per day)			
		Mini-mum	Mean	Median	Maxi-mum
Overburden	72	1.3	1,694	161	29,310
Basalt units	573	.08	805	70	58,370
Older bedrock	14	.12	46	6	258

and median values for the Overburden unit were 1,694 and 161 ft/d, respectively. The specific-capacity derived mean and median values for the basalt units were 805 and 70 ft/d, respectively, and for the bedrock units 46 and 6 ft/d, respectively. The median values are similar in magnitude to values reported by Freeze and Cherry (1979) for similar materials and to values described previously. The information in table 3 also indicates that, overall, the values for the overburden and CRBG units are similar. Note that the bedrock values are based on a limited number of tests, and data for wells identified as 'dry' or little production were not included, resulting in a likely bias to larger values. The minimum values illustrate that zones of low K_h are present in most units, and the large range in K_h indicates substantial heterogeneity in the units.

Vertical Hydraulic Conductivity

Vertical hydraulic conductivity (K_v) is an important hydraulic characteristic that is difficult to measure. It is a measure of a material's ability to transmit water vertically (the impedance to downward/upward flow) and is expressed in units of feet per day. K_v is a major control on the movement of water in the CPRAS flow system—lateral and vertical variations in K_v affect vertical hydraulic gradients, and therefore, flow rates into, within, and out of units. Except for the work of Drost and others (1997), who used measured canal seepage rates to calculate K_v, nearly all of the previous estimates of K_v were derived using groundwater-flow models. Previous estimates of vertical hydraulic conductivity within the CPRAS are summarized in the following sections and listed in table 4.

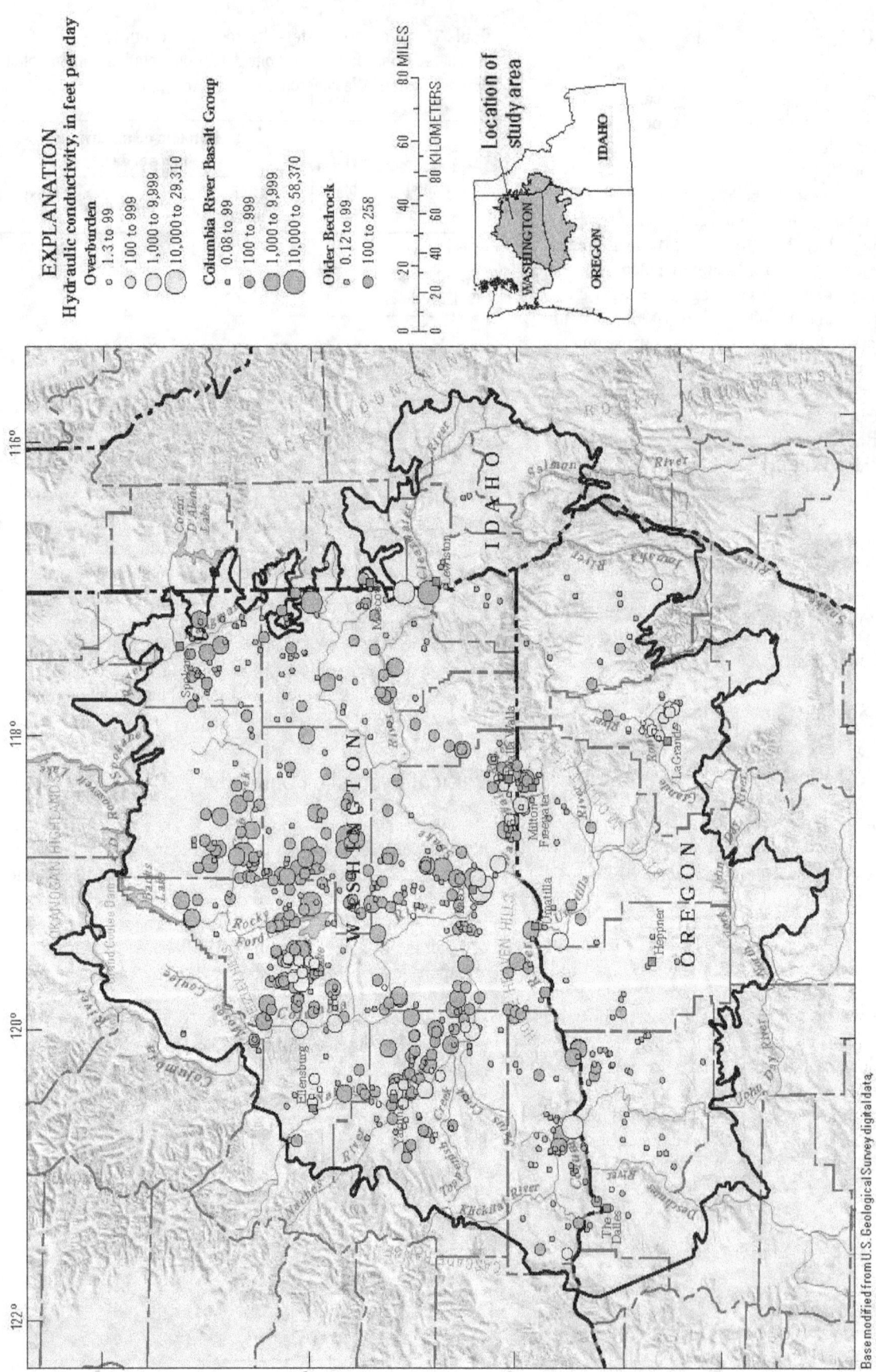

Base modified from U.S. Geological Survey digital data.

Figure 8.   Distribution of estimated horizontal hydraulic conductivity in the Columbia Plateau Regional Aquifer System, Washington, Oregon, and Idaho.

Table 4.   Summary of previous estimates of vertical hydraulic conductivity, Columbia Plateau Regional Aquifer System, Washington, Oregon, and Idaho.

[Abbreviations: –, not applicable]

Unit	Basin	Vertical hydraulic conductivity (feet per day)				Source
		Minimum	Median	Mean	Maximum	
Overburden Unit						
Loess	Columbia Plateau	–	–	0.05	–	Smoot and Ralston, 1987; Lum and others, 1990
Young alluvium	Lower Satus Creek	–	–	0.13	–	Prych, 1983
Alluvial aquifer	Toppenish	–	–	0.009	–	Bolke and Skrivan, 1981; U.S. Geological Survey, 1975
Slack water deposits (Touchet beds)	Pasco	–	–	0.4	–	Drost and others, 1997
Glaciofluvial deposits (Pasco gravel)	Pasco	–	–	0.7	–	Drost and others, 1997
Overburden	Columbia Plateau Aquifer System	4.0E-7	2.0		1,400	Hansen and others, 1994
Ellensburg Formation, upper (old alluvium)	Lower Satus Creek	–	–	0.09	–	Prych, 1983
Ringold Formation, upper	Pasco	–	–	0.4	–	Drost and others, 1997
Ringold Formation	Benton Basin	–	–	0.009	–	Zimmerman, 1983
Basalt Units						
Saddle Mountains	Pasco	–	–	0.3	–	Drost and others, 1997
Saddle Mountains and Wanapum	Umatilla Basin	4.0E-7	–	–	2.0E-4	Davies-Smith and others, 1988
Saddle Mountains and Wanapum, anticline zone	Horse Heaven Hills	–	–	5.0E-4	–	Packard and others (1996)
Saddle Mountains and Wanapum, syncline zone	Horse Heaven Hills	–	–	0.01	–	Packard and others (1996)
Wanapum	Pullman-Moscow Basin	8.0E-4	–	–	1.2E-3	Smoot and Ralston, 1987; Lum and others, 1990
Grande Ronde	Pullman-Moscow Basin	1.0E-4	–	–	2.5E-3	Smoot and Ralston, 1987; Lum and others, 1990
Basalt	Columbia Plateau Aquifer System	5.0E-5	0.001	–	7.0	Hansen and others, 1994
Basalt	Columbia Plateau Aquifer System	5.0E-4	–	–	4.0	Whiteman and others, 1994
Basalt	Mosier	1.0E-6	–	–	0.001	Burns, 2011, U.S. Geological Survey, written commun.
Dense basalt control (base case)						
Saddle Mountain						
Undisturbed	Pasco	–	–	0.001	–	Zimmerman, 1983
Anticline	Pasco	–	–	0.009	–	Zimmerman, 1983
Fault	Pasco	–	–	0.009	–	Zimmerman, 1983
Wanapum						
Undisturbed	Pasco	–	–	0.008	–	Zimmerman, 1983
Anticline	Pasco	–	–	0.03	–	Zimmerman, 1983
Fault	Pasco	–	–	0.006	–	Zimmerman, 1983
Barrier	Pasco	–	–	0.05	–	Zimmerman, 1983
Grande Ronde		–	–	–	–	Zimmerman, 1983
Undisturbed	Pasco	–	–	3.0E-5	–	Zimmerman, 1983
Anticline	Pasco	–	–	4.0E-4	–	Zimmerman, 1983
Fault	Pasco	–	–	9.0E-5	–	Zimmerman, 1983
Barrier	Pasco	–	–	0.002	–	Zimmerman, 1983

Table 4.   Summary of previous estimates of vertical hydraulic conductivity, Columbia Plateau Regional Aquifer System, Washington, Oregon, and Idaho.—Continued

[Abbreviations: ft, foot; –, not applicable]

Unit	Basin	Vertical hydraulic conductivity (feet per day)				Source
		Minimum	Median	Mean	Maximum	
Interbed basalt control (alternate base case)						
Saddle Mountain						
Undisturbed	Pasco	–	–	0.001	–	Zimmerman, 1983
Anticline	Pasco	–	–	1.0E-5	–	Zimmerman, 1983
Fault	Pasco	–	–	0.009	–	Zimmerman, 1983
Wanapum						Zimmerman, 1983
Undisturbed	Pasco	–	–	0.008	–	Zimmerman, 1983
Anticline	Pasco	–	–	1.0E-4	–	Zimmerman, 1983
Fault	Pasco	–	–	0.006	–	Zimmerman, 1983
Barrier	Pasco	–	–	0.05	–	Zimmerman, 1983
Grande Ronde						Zimmerman, 1983
Undisturbed	Pasco	–	–	3.0E-5	–	Zimmerman, 1983
Anticline	Pasco	–	–	4.0E-4	–	Zimmerman, 1983
Fault	Pasco	–	–	9.0E-5	–	Zimmerman, 1983
Barrier	Pasco	–	–	0.002	–	Zimmerman, 1983

Overburden Unit

Large differences in types of sedimentary deposits results in large variations in estimates of K_v. Hansen and others (1994) estimated K_v ranging from 4×10^{-7} to almost 1,400 ft/d for the Overburden unit throughout the Columbia Plateau. The median value of K_v was 2 ft/d, and the average vertical anisotropy was 25:1. On the basis of the previously described K_h values, effective regional K_v values for the deposits comprising the Overburden unit likely range from 0.1 to 1 ft/d.

Smoot and Ralston (1987) and Lum and others (1990) estimated a value of 0.05 ft/d for loess on the basis of a vertical anisotropy of 100:1. For young alluvium in the lower Satus Creek Basin, Prych (1983) estimated a mean K_v of 0.13 ft/d. Bolke and Skrivan (1981) assigned a constant K_v of 0.009 ft/d in a groundwater model of the Toppenish alluvial aquifer. For the lower Satus Creek Basin, Prych (1983) assumed a K_h:K_v anisotropy ratio of 1,000:1 for the old alluvium and upper Ellensburg Formation, yielding a K_v of 0.09 ft/d in a groundwater model.

Drost and others (1997) used irrigation canal-seepage rates to estimate K_v in the Pasco Basin. This method resulted in mean values of 0.4, 0.7, and 0.4 ft/d for the Touchet Beds, Pasco gravels, and upper Ringold Formation, respectively. For the sedimentary material overlying the Saddle Mountains unit in the Eastern Benton Basin, Zimmerman (1983) assumed K_v to be controlled by the clays of the Ringold Formation and used a constant value 0.009 ft/d in a groundwater-flow model.

The K_v of the fine-grained (such as claystone, mudstone, and shale) parts of the Overburden unit also is unknown but likely is as small as 10^{-10} to 10^{-6} ft/d assuming a K_h:K_v ratio

of 1,000:1. Bredehoeft and others (1983) estimated K_v of the Pierre Shale and Cretaceous shale in South Dakota to be about 4×10^{-6} and 5×10^{-4} ft/d, respectively. They also noted that values greater than 4×10^{-6} for the thick Cretaceous shale would not support an underlying, flowing-artesian aquifer system owing to increased vertical leakage. That is, the known upward vertical leakage through the shale has a threshold that is limited by the value of K_v. For the thick upper part of the late-Cretaceous Hell Creek Formation (shale dominated) in Montana, Hotchkiss and Levings (1986) used aquifer-test data to estimate a maximum range in K_v from 5×10^{-5} to 0.001 ft/d. Model-derived K_v values for the shale-dominated confining units in a 42,000 mi^2 area of eastern Montana and northeastern Wyoming were about 2×10^{-5} ft/d (Hotchkiss and Levings, 1986). Taken together, the summarized information indicates that the K_v of thick shale within the Overburden unit probably ranges from about 10^{-6} to 10^{-3} ft/d.

Columbia River Basalt Group Units

Zimmerman (1983) examined two possible controlling mechanisms of K_v for different basalt zones. In the "base case," dense basalts were assumed to be the primary control on vertical flow. In the "alternate base case," interbeds were the primary control. In both cases, dense basalts were assumed to control vertical flow in undisturbed zones. Zimmerman assigned vertical anisotropy ratios (K_h:K_v) of 1,000:1 to all groundwater-model nodes in undisturbed zones, which yielded K_v values ranging from 3×10^{-5} to 0.001 ft/d for the dense basalts. Fault and barrier zones for both cases were simulated as isotropic (K_h:K_v = 1), which may be similar to the anisotropy of a basalt flow interior. Differences between the two cases occurred along anticlines where interbeds rather than dense basalts were assumed to control vertical flow. Davies-Smith and others (1988) also estimated that the Mabton unit, in contrast to basalt, exerted more control on the vertical movement of water in the Umatilla Basin in Oregon, where the Mabton is composed of fine-grained deposits. Vertical anisotropy for Saddle Mountains and Wanapum were assigned values of 8,333:1 and 2,063:1, respectively, yielding values for K_v ranging from 4×10^{-7} to 2×10^{-4} ft/d (Davies-Smith and others, 1988).

Vertical anisotropy for the Wanapum unit was estimated to be 500:1, with K_v ranging from 8.0×10^{-4} to 1.2×10^{-3} ft/d in the Pullman-Moscow Basin in Washington and Idaho (Smoot and Ralston, 1987; Lum and others, 1990). These investigations estimated the Grande Ronde unit anisotropy to range from 5,000:1 to 2,000:1, with K_v ranging from 1×10^{-4} to 2.5×10^{-3} ft/d.

Whiteman and others (1994) reported that K_v of the CPRAS was largely unknown, but estimated that values ranged from 5×10^{-4} to 4 ft/d, with vertical anisotropy of 1,000:1 to 100:1. Hansen and others (1994) estimated that K_v in the CPRAS ranged from 5×10^5 to 7 ft/d, with a median value of 1×10^{-3} ft/d. Typical vertical anisotropy was 1,500:1 to 1,000:1.

Packard and others (1996) estimated K_v for two zones (anticlines and synclines) in a groundwater-flow model of the Horse Heaven Hills on the southeastern border of the basin. A K_v value of 5×10^{-4} ft/d was calibrated for the anticline zone for the Saddle Mountains unit and Wanapum unit, whereas along the synclines the average value was larger (0.01 ft/d). Drost and others (1997) estimated K_v to be 0.3 ft/d for the Saddle Mountains unit on the basis of canal-seepage losses. Although this K_v is larger than most other reported values, it was based on water-level changes with time resulting from a known amount of seepage (recharge).

Older Bedrock Unit

The authors have been unable to locate previous estimates of K_v for the bedrock in the study area. K_v values for these units likely range over several orders of magnitude and vary by the type of materials comprising a unit, for example, schist in contrast to sandstone. For the Older Bedrock unit as a whole, the average vertical anisotropy would be large, perhaps on the order of 10,000:1 to 2,000:1. On the basis of information described above, K_v for the shale/clay parts of the sedimentary units may range from 4×10^{-6} to $\times 10^{-4}$ ft/d.

Storage Coefficient

The storage coefficient is a measure of a unit's ability to store and release water and is defined as the volume of water that a unit will absorb or release from storage per unit surface area per unit change in head. It is expressed in units of cubic feet per cubic feet, a dimensionless quantity. Storage coefficients for a confined aquifer can range from 5×10^{-5} to 0.005; values for an unconfined aquifer (referred to as specific yield) are much larger and can range from 0.01 to 0.30 (Freeze and Cherry, 1979). Previous estimates of storage coefficients within the CPRAS are summarized in the following sections and listed in table 5.

Analyses of aquifer tests in the Pasco Basin yielded unconfined storage-coefficient values of 0.1 for the middle Ringold Formation and 0.15 to 0.2 for Pasco gravels (Drost and others, 1997). Aquifer tests in the confined parts of the Pasco Basin yielded storage coefficients of 0.03 to 0.07 for Pasco gravels, 7×10^{-5} to 0.06 for the middle Ringold Formation, 0.002 to 0.05 for the lower Ringold Formation, and 1×10^{-6} to 0.006 for the CRBG (Drost and others, 1997). Drost and others (1997) also reported that previous modeling studies calibrated CRBG storage-coefficient values of 0.0001 to 0.01. Vermeul and others (2001) estimated storage coefficients for the coarse-grained sediments of the unconfined parts of the Hanford and Ringold Formations as 0.07 and 0.2, respectively. Model-derived storage coefficients for all confined units were 1×10^{-6}.

On the basis of grain size, Drost and others (1997) estimated storage coefficients for the Touchet Beds (0.08), upper Ringold Formation (0.07 to 0.2), and lower Ringold Formation (0.02 to 0.2). Prych (1983) used an unconfined storage coefficient of 0.1 for the Touchet Beds in the lower Satus Creek Basin. An unconfined storage coefficient for the Toppenish alluvial aquifer was estimated at 0.2 (U.S. Geological Survey, 1975).

Golder Associates (2002) estimated a storage coefficient of 2×10^{-3} for the lower part of the upper Ellensburg Formation on the basis of buildup and drawdown data during aquifer tests. Converse Consultants NW (1991) used data from a 24-hour aquifer test to estimate a storage coefficient of about 7×10^{-4} for the lower part of the upper Ellensburg Formation.

Whiteman and others (1994) indirectly estimated storage coefficients for the CRBG units on the basis of specific storage. Median values for the Saddle Mountains, Wanapum, and Grande Ronde units were 2×10^{-5}, 3×10^{-5}, and 2×10^{-4}, respectively. For the Overburden unit, they estimated values ranging from 2×10^{-4} to 0.2.

Previously estimated storage coefficients were initially used by Hansen and others (1994, table 4) in the CPRAS model. Model-derived specific yields for the overburden aquifer ranged from 0.1 to 0.2. The estimated median storage coefficients for the Saddle Mountains, Wanapum, and Grande Ronde were 4×10^{-5}, 4×10^{-5}, and 2×10^{-4}, respectively. Packard and others (1996) initially assigned CRBG storage coefficients on the basis of previous studies in a groundwater model of the Horse Heaven Hills. Transient calibration of the model yielded values of 0.001 for the Grande Ronde and Wanapum and 0.01 for the Saddle Mountains.

Table 5.   Summary of previous estimates of storage coefficients, Columbia Plateau Regional Aquifer System, Washington, Oregon, and Idaho.

[Abbreviations: -, not applicable]

Unit	Storage coefficient		Source
	Range	Median	
Alluvium (Toppenish)	–	0.2	U.S. Geologial Survey, 1975
Overburden aquifers	2.0E-04 to 0.2	–	Whiteman and others, 1994
Overburden aquifers	0.1 to 0.2	–	Hansen and others, 1994
Slack water deposits (Touchet beds)	–	0.08	Drost and others, 1997
Slack water deposits (Touchet beds)	–	0.1	Prych, 1983
Coarse-grained deposits of the Hanford Formation (unconfined)	–	0.07	Vermeul and others, 2001
Coarse-grained deposits of the Ringold Formation (unconfined)	–	0.2	Vermeul and others, 2001
Glaciofluvial deposits - Pasco gravel (unconfined)	0.15 to 0.2	–	Drost and others, 1997
Glaciofluvial deposits - Pasco gravel (confined)	0.03 to 0.07	–	Drost and others, 1997
Ellensburg Formation, upper	–	2.0E-3	Golder, 2002
Ellensburg Formation, upper	–	7.0E-4	Converse Consultants NW, 1991
Ringold Formation, upper (based on grain size)	0.7 to 0.2	–	Drost and others, 1997
Ringold Formation, middle (unconfined)	–	0.1	Drost and others, 1997
Ringold Formation, middle (confined)	7.0E-05 to 0.06	–	Drost and others, 1997
Ringold Formation, lower (based on grain size)	0.02 to 0.2	–	Drost and others, 1997
Ringold Formation, lower (confined)	0.002 to 0.05	–	Drost and others, 1997
Saddle Mountains	1.2E-06 to 7.8E-05	2.0E-5	Whiteman and others, 1994
Saddle Mountains	3.7E-06 to 1.1E-04	4.0E-5	Hansen and others, 1994
Saddle Mountains (Horse Heaven Hills)	–	0.01	Packard and others, 1996
Wanapum	1.8E-06 to 9.9E-05	3.0E-5	Whiteman and others, 1994
Wanapum	3.0E-06 to 2.3E-04	4.0E-5	Hansen and others, 1994
Grande Ronde	6.0E-06 to 1.1E-03	2.0E-4	Whiteman and others, 1994
Grande Ronde	1.2E-05 to 1.1E-03	2.0E-4	Hansen and others, 1994
Grande Ronde and Wanapum (Horse Heaven Hills)	–	0.001	Packard and others, 1996
Basalt (confined)	1.0E-06 to 0.006	–	Drost and others, 1997
Basalt	0.01 to 1.0E-04	–	Smoot and Ralston, 1987; and Davies-Smith and others, 1988

Groundwater Occurrence and Movement

The occurrence and movement of groundwater in the CPRAS was studied in detail in the 1980s and is described in several reports including Bauer and others (1985), Whiteman (1986), and Lane and Whiteman (1989). A general discussion of the occurrence and movement of water is included in Hanson and others (1994) and Vaccaro (1999). Generalized groundwater-elevation maps of the Overburden unit and the Saddle Mountains, Wanapum, and Grande Ronde Basalt units from spring 2009 are provided in Snyder and others (2010). Detailed descriptions of the occurrence and movement of water in the Yakima River Basin is included in Vaccaro and others (2009). The following sections provide an overview of the groundwater occurrence and movement in the CPRAS summarized from the references listed at the beginning of this section.

Groundwater Occurrence

Groundwater in the basalts occurs in joints, vesicles, fractures, and other local features that create permeable zones, and in intergranular pores of the sedimentary interbeds (Hanson and others, 1994). High permeabilities generally occur in flow tops; in vesicular zones at the base of colonnades; and in basal, pillow-palagonite complexes (fig. 5). The more dense and coherent entablature and most of the colonnade probably have low permeabilities. Sharp folding and faulting can cause shearing and fracturing of the basalt flows and create local areas with large permeabilities in joints and fractures, whereas shear faulting can offset the interflow zones and disrupt their hydraulic continuity. Newcomb (1969) reported that in the Dalles area, fault zones block the movement of groundwater by destruction and offset of permeable zones. Displacement of individual flows along faults, however, can locally enhance vertical movement of water by providing fractured zones across basalt flows that could serve as conduits for vertical groundwater flow (Reidel and others, 2002). Basalt permeability undoubtedly has increased in places where erosion has beveled and dissected basalt units. It appears that a combination of the permeability increases described above result in the highest primary permeabilities in the basalts. Local increases in the thickness of sedimentary interbeds result in increased storage capacity of the system at that interval and also alter the permeability of the system. In some areas, sedimentary interbeds serve as aquifers where their lithologies facilitate the storage and transmission of water.

Groundwater-Flow System

Groundwater moves through the regional aquifer system from the uplands (high land-surface altitude—topographic highs) to surface drainage features in the lowlands, principally to the Columbia River and its principal tributaries—such as the Snake, Yakima, John Day, Umatilla, Spokane, Klickitat, and Deschutes Rivers. Groundwater movement is affected by topography, geologic structure, natural recharge, discharge locations, hydraulic characteristics, recharge from the use of water (principally surface-water irrigation), and groundwater pumpage.

Within the basalt units, groundwater moves both laterally and vertically in the basalt-interflow zones, flow centers, and sedimentary interbeds. Lateral hydraulic conductivities generally are assumed to be greatest in the interflow zones primarily because of features such as flow breccia, rubble, and vesicles. Hydraulic conductivities in the flow centers are controlled by secondary features—the predominantly vertical joints and fractures of the entablature and colonnade. Consequently, the interflow zones support most of the lateral groundwater movement, whereas movement in the flow center is mainly vertical. Therefore, the interflow zones in the basalt sequence are numerous, thin, semiconfined aquifers whose physical and hydraulic characteristics vary laterally and vertically.

Except for groundwater flow in the deeply buried parts of the system, large-scale structural control compartmentalizes the flow system in places (Kinnison and Sceva, 1963; Hansen and others, 1994; Bauer and Hansen, 2000; Vaccaro and others, 2009). The compartmentalization limits the length of the flow paths, resulting in relatively short paths for such a large aquifer system. Structural control is exerted primarily by the major ridges in the Yakima Fold Belt (Hansen and others, 1994; Reidel and others, 2002; Jones and others, 2006).

Groundwater levels in the basalt units generally parallel the land surface or, where a unit is buried, parallel the dip of the basalt because most groundwater occurs and moves in the interflow zones. Where the units are deeply buried, water-level contours are smoother (lower hydraulic gradient) than those in the uplands, which typically are outcrop areas. Water-level contours mapped for the CRBG units are generalized owing to sparse data in many locations and large variations in depth of water levels. Additionally, spatial and vertical variations in hydraulic characteristics of both individual flows and interbedded sediments as well as the presence of geologic structure result in a much more complex flow system for each unit than is depicted on generalized groundwater-elevation maps.

The potentiometric surface of the Saddle Mountains unit generally parallels the land surface in areas where little or no overburden is present. Groundwater flows toward surface-drainage features; this pattern of flow is similar in the Wanapum and Grande Ronde units where they are not overlain by another unit. Flow in the deeply buried parts of the Wanapum and Grande Ronde units is less controlled by surface-drainage features. Instead, water movement is controlled by vertical leakage. For example, in the Quincy Basin, groundwater within the Wanapum unit, which is at or near land surface, flows toward Moses Lake and Potholes Reservoir. Groundwater in the underlying Grande Ronde unit (especially its deeper parts), however, is relatively unaffected by the local surface-water features and flows south toward the Columbia River.

Groundwater flow in the Grande Ronde unit also is compartmentalized but not to the same extent as in either the Saddle Mountains or Wanapum units. The large spatial extent of the Grande Ronde unit results in a large flow system with more interconnections than in the other two CRBG units. Where the unit crops out, the water-level contours mimic land-surface topography and they become a more subdued replica of topography as the unit becomes buried. In the more deeply buried parts of the unit, the contours are smoother than those for the other CRBG units. Similarly, its water-level contours near geologic structures in the eastern part of the area are more subdued and smoother. The flow system in the Grande Ronde unit is controlled by the regional discharge locations along the Columbia River and major tributaries; that is, the regional flow (hydraulic head) in the Grande Ronde unit tends to the level of the major streams (fig. 9). There may be a regional flow system in the deeper part of the unit but there are insufficient data to verify the presence of such a system.

Within the Palouse subprovince, north of the Snake River, groundwater in both the Wanapum and Grande Ronde units flows toward the southwest, and regional discharge is to the Columbia and Snake Rivers. Water levels closely parallel the land surface and the regional slope of the basalts. In the Yakima Fold Belt, groundwater flows downward from the anticlinal ridges toward streams and rivers in the intervening synclinal basins.

Groundwater flow in the Overburden unit is similar to that in the outcrop areas of the basalts. Recharge mainly is from precipitation and infiltration of irrigation water. Discharge is to rivers, lakes, drains and waterways, wells, and to the underlying basalt unit. Downward movement of water to the basalts is controlled by the vertical conductivity, unit thickness, and head differences between the units.

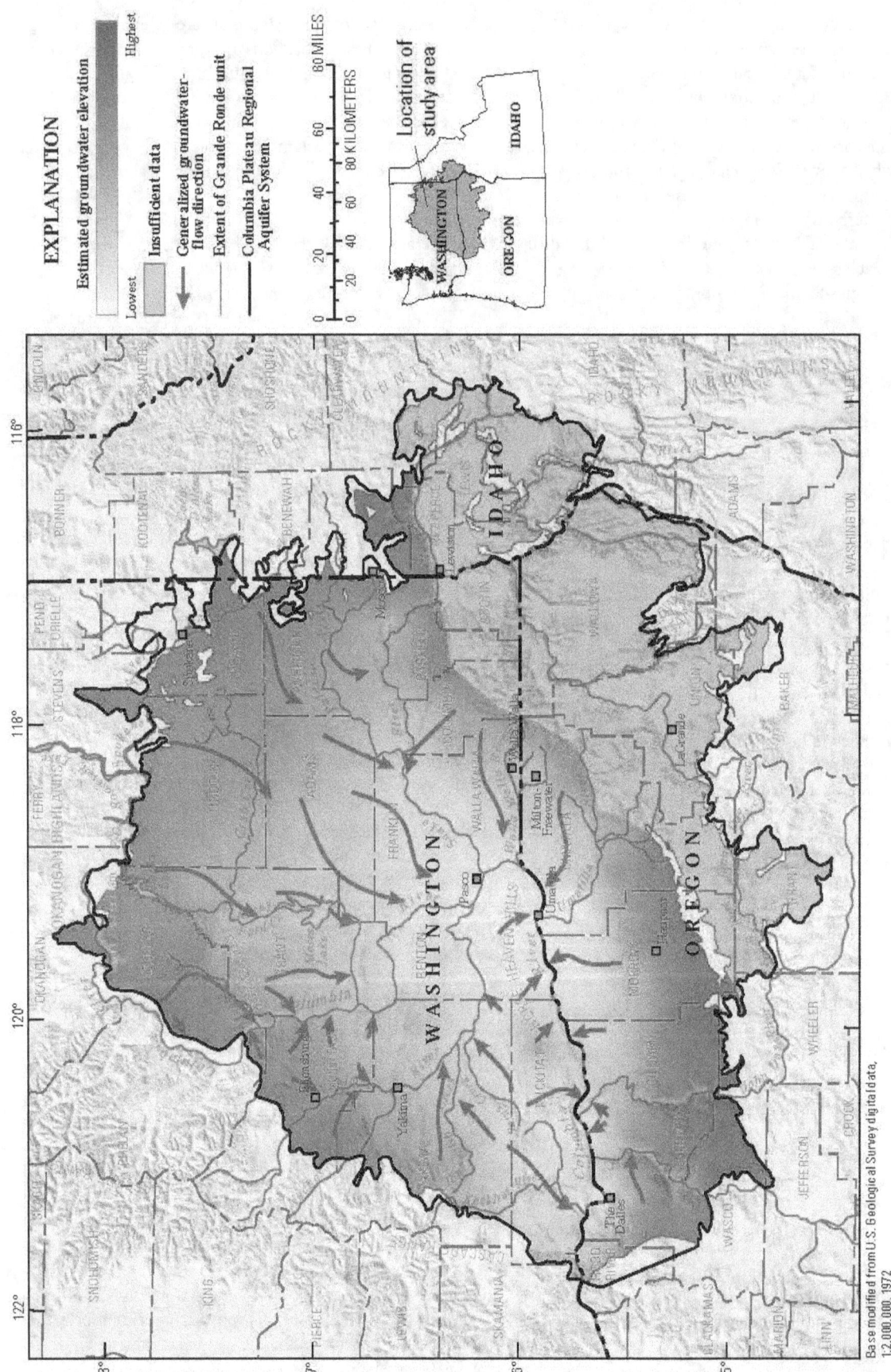

Base modified from U.S. Geological Survey digital data,
1:2,000,000, 1972

Figure 9.   Generalized groundwater levels and directions of lateral groundwater movement for the Grande Ronde unit, Columbia Plateau Regional Aquifer System, Washington, Oregon, and Idaho (modified from Vaccaro, 1999 and Snyder and others, 2010).

Groundwater Age

An assessment of existing isotopic data and published literature was done to better understand groundwater age in the CPRAS. Overall patterns of Holocene and Pleistocene groundwater occurrence are described herein, based primarily on analysis of carbon-14 and oxygen-18 data. These patterns of groundwater age place constraints on groundwater movement and could be used as general calibration constraints in flow-model particle-tracking calibration.

Carbon-14 (^{14}C), a radioactive isotope of carbon, is naturally present in Earth's atmosphere, and hence, Earth's biosphere. Atmospheric carbon dioxide (CO_2), containing ^{14}C and other isotopes of carbon, is partitioned into water that is in contact with the atmosphere, thus becoming dissolved inorganic carbon (DIC). Additional DIC in unsaturated-zone water is derived from biological activity: microbially mediated oxidation of organic carbon and root respiration. Through these processes, unsaturated zone water and shallow groundwater obtain ^{14}C.

Concentrations of ^{14}C in groundwater generally decrease as groundwater moves through an aquifer. These decreases occur in response to two processes: radioactive decay of ^{14}C and dilution of ^{14}C. Radioactive decay of ^{14}C, which has a half-life of 5,730 years, is the time-dependent process that is exploited for age-dating. Changes in ^{14}C concentration along flow paths can be used to estimate groundwater ages of up to several tens of thousands of years by comparing ^{14}C concentrations measured in groundwater samples to original (that is, beginning-of-flow path) ^{14}C concentrations. The concentration of ^{14}C present at the beginning of a flow path sometimes is taken to be about 100 percent modern carbon (PMC) (the atmospheric concentration present in 1950, before widespread thermonuclear weapons testing increased atmospheric concentrations of ^{14}C). Age-dating with ^{14}C might be straightforward were it not for the potential for ^{14}C dilution to occur along groundwater flow paths, a process by which DIC is diluted with ^{14}C-free (so-called "^{14}C-dead") carbon. Common sources of ^{14}C-dead carbon in aquifers include sedimentary organic material and calcite (including pedogenic calcite in basalt-interflow zones and hydrothermally emplaced calcite); magmatic CO_2 also has been identified as a source of ^{14}C-dead carbon to groundwater. Exchange of carbon atoms between aqueous and solid phases, and diffusion of ^{14}C into aquitards, also can lead to dilution of original ^{14}C in DIC.

Oxygen-18 (^{18}O) and oxygen-16 (^{16}O) are stable isotopes of oxygen. ^{18}O and ^{16}O are the dominant oxygen isotopes in water molecules. The amount of ^{18}O in water is expressed by comparing the ratio of ^{18}O to ^{16}O in a water sample to the ratio of ^{18}O to ^{16}O in the reference standard Vienna Standard Mean Ocean Water (Coplen, 1994), and reported using delta (δ) notation and units of per mil (‰). Oxygen isotopes are a particularly valuable tool that can complement ^{14}C data.

Several environmental processes, such as evaporation, condensation, chemical reaction, and diffusion, affect $\delta^{18}O$ values. As a result, $\delta^{18}O$ values of groundwater recharge that are derived from atmospheric precipitation vary in space and time. Higher-elevation recharge often is isotopically lighter (lower $\delta^{18}O$ values) than lower-elevation recharge from sites that are otherwise similar (for example, Vaccaro and others, 2009), and recharge in zones located downwind from atmospheric moisture sources often is isotopically lighter than that found in zones located closer to those atmospheric moisture sources (Kendall and Coplen, 2001). Recharge that occurred during the Pleistocene generally was isotopically lighter than recharge that fell during the Holocene for the same geographic area (Clark and Fritz, 1997, p. 198–200). Differences in the stable-isotope composition of oxygen (as well as hydrogen) atoms in water molecules have been widely exploited in the hydrologic and climate-change literature.

Selected existing ^{14}C and $\delta^{18}O$ data were compiled to demonstrate some of the general patterns of groundwater-age structure that are found in the CPRAS. This analysis represents a highly simplified analysis of a complex topic. Focused analysis of geochemical and isotopic data for selected zones within the larger CPRAS can be found in the hydrologic literature, but the analysis presented herein is a more generalized contribution.

Two large sets of ^{14}C and $\delta^{18}O$ data providing wide spatial coverage were identified for this effort. One dataset was a compilation of CPRAS data collected in the 1980s (Wagner and Lane, 1994). The other dataset was a compilation of data collected in the Yakima River Basin between 1990 and 2005 (Vaccaro and others, 2009). A total of 100 sets of samples from 84 sites contained both ^{14}C concentrations and $\delta^{18}O$ values, as well as locational data. The locations of these sites are shown in figure 10. Open intervals ranged from 0 to 2,715 ft below land surface. A comparison of ^{14}C concentrations and $\delta^{18}O$ values is shown in figure 11. Concentrations of ^{14}C range from greater than 100 PMC, likely representing contributions of ^{14}C from atmospheric testing of thermonuclear weapons ("bomb" ^{14}C), to less than 2 PMC, a concentration that could equate to a travel time of greater than 30,000 years if it could be assumed that the ^{14}C concentration in recharge was 100 PMC and if the decrease in ^{14}C concentrations could be attributed solely to radioactive decay.

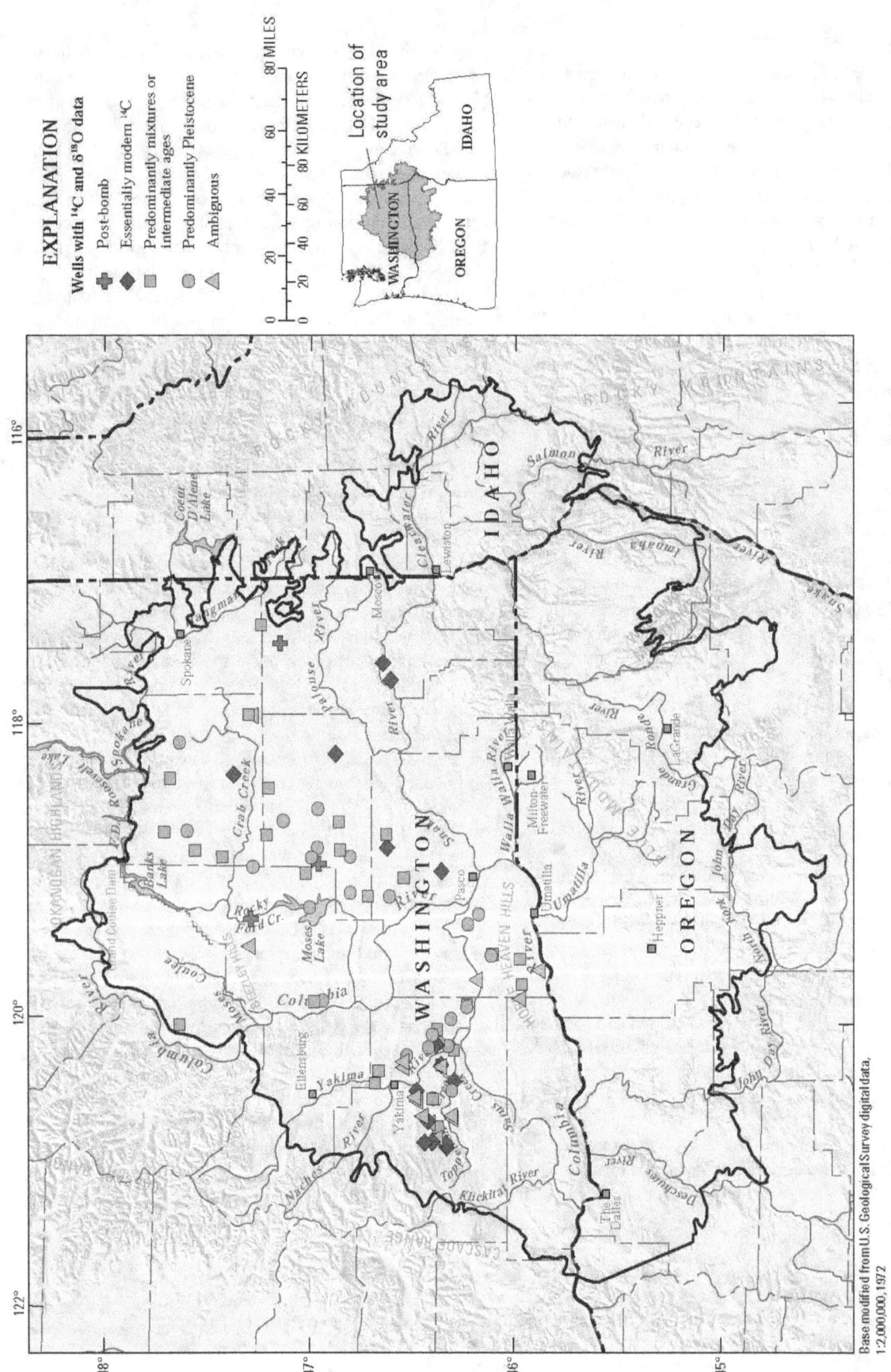

Base modified from U.S. Geological Survey digital data,
1:2,000,000, 1972

Figure 10.   Locations of wells with carbon-14 (^{14}C) and delta oxygen-18 (δ^{18}O) data compiled for analysis in this report. [Symbols indicate interpreted categories as explained in the text and indicated in figure 11.]

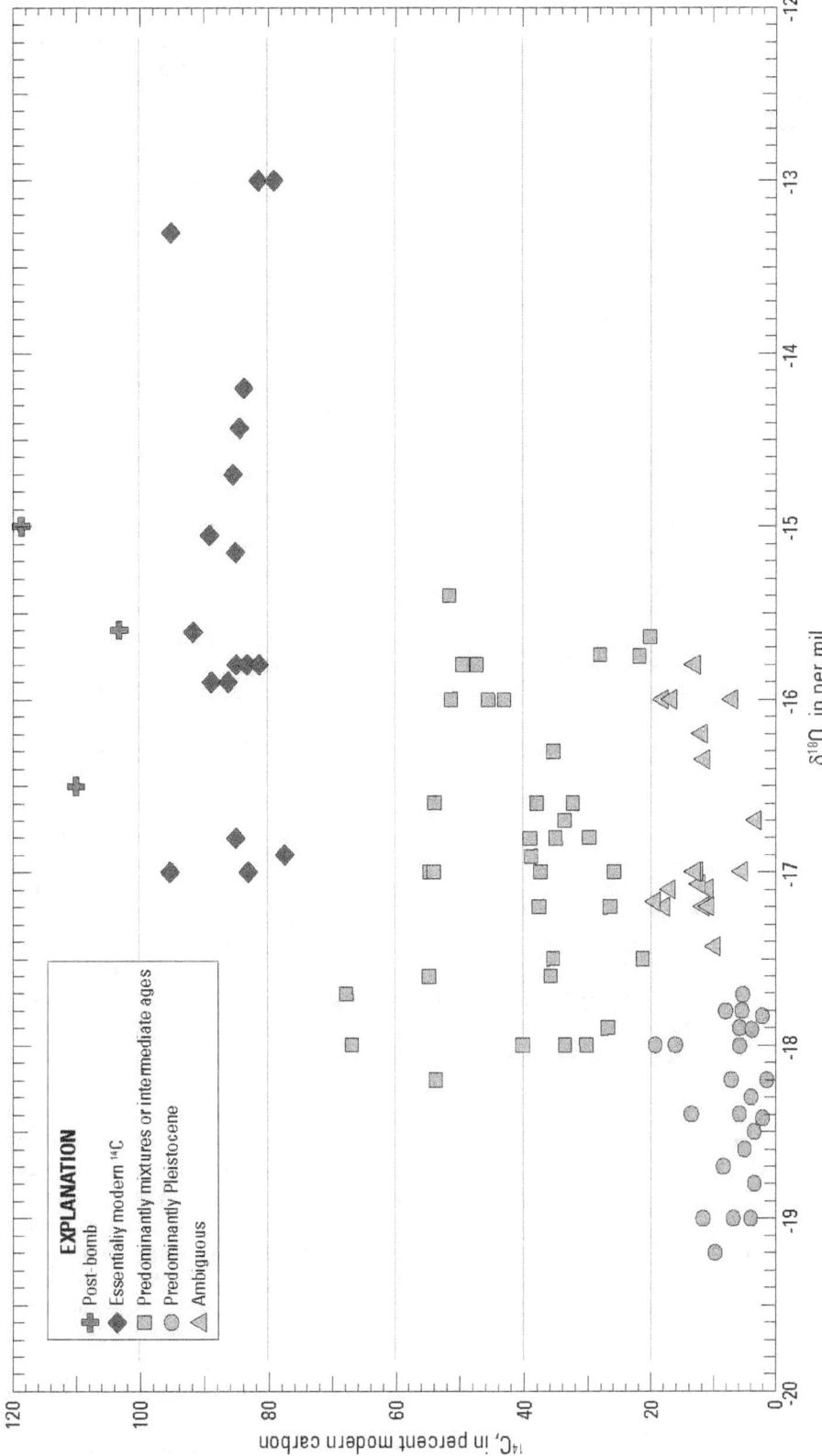

Figure 11. Relation between delta oxygen-18 ($\delta^{18}O$) values and carbon-14 (^{14}C) concentrations, Columbia Plateau Regional Aquifer System.

Initial ^{14}C concentrations in basalt-aquifer recharge, although sometimes assumed to be 100 PMC (for example, Douglas and others, 2007, and Brown and others, 2010), probably were diluted by ^{14}C-dead carbon in the recharge zone, yielding initial ^{14}C concentrations of less than 100 PMC except where bomb ^{14}C was present. Analysis of ^{14}C content of shallow groundwater recharge in CRBG aquifers has yielded estimates of initial ^{14}C content of around 80 to 85 PMC (Yakima River Basin; Vaccaro and others, 2009) and approximately 75 PMC (U.S. Department of Energy Hanford Site; Reidel and others, 2002) for groundwater unaffected by anthropogenic activities. These estimates of initial ^{14}C concentrations reflect slight dilution of initial ^{14}C with ^{14}C-dead or low-^{14}C carbon in the unsaturated zone or shallow groundwater system.

Carbon mass transfers along groundwater flow paths can dilute the ^{14}C content of groundwater. It has been argued that the low-carbon content of basalt aquifers results in negligible carbon mass transfers (for example, Douglas and others, 2007; Brown and others, 2010). Others have argued that these aquifers may contain sedimentary organic material in interflow zones or pedogenic or hydrothermal calcite that can substantially affect ^{14}C age-dates (Hinkle, 1996; Steinkampf and Hearn, 1996). Vaccaro and others (2009) noted a general pattern of increasing delta carbon-13 ($\delta^{13}C$) values with decreasing ^{14}C concentrations in CRBG aquifers. This was interpreted to indicate additions along groundwater flow paths of carbon enriched in ^{13}C and depleted in ^{14}C, possibly associated with methanogenesis or calcite dissolution (Vaccaro and others, 2009). Although methanogenesis and calcite dissolution are reasonable hypotheses to explain the observed relation between $\delta^{13}C$ values and ^{14}C concentrations, other processes could explain the ^{13}C enrichment observed by Vaccaro and others (2009). For example, Blaser and others (2010) observed a similar relation between $\delta^{13}C$ values and ^{14}C concentrations (albeit for an aquifer in Belgium). Blaser and others (2010) presented noble gas and isotope evidence indicating that differences in climate between the Pleistocene and Holocene changed the balance of atmospheric and biological soil-zone CO_2 in recharge and thus $\delta^{13}C$ values of DIC in recharge. These and other processes that affect the concentrations and isotopic character of DIC introduce uncertainties in ^{14}C dating, and detailed analysis and multiple lines of evidence often are required to develop defensible ^{14}C-based groundwater ages. However, general patterns of the overall age structure of groundwater in a regional aquifer system often can be elucidated with ^{14}C and other data, and such understanding of groundwater age can be useful for characterizing directions and approximate timescales of groundwater flow.

The ^{14}C and $\delta^{18}O$ data shown in figure 11 appear to fall into several groups. One group is a small group of post-bomb samples—those with ^{14}C concentrations greater than 100 PMC. These samples represent modern (primarily post-bomb) recharge. $\delta^{18}O$ values are greater than (more positive than) -17.0‰. These groundwater samples were from shallow wells.

Samples with ^{14}C concentrations ranging from 75 to 100 PMC represent another group. These samples contain ^{14}C concentrations essentially representative of modern water (Reidel and others, 2002; Vaccaro and others, 2009); $\delta^{18}O$ values range from -17.0 to -13.0‰.

Samples with ^{14}C concentrations of less than 20 PMC could be consistent with recharge during the Pleistocene. Given an initial ^{14}C concentration of 80 PMC and assuming no carbon mass transfers, a ^{14}C concentration of less than 20 PMC equates to a ^{14}C age of greater than 11,000 years. However, carbon mass transfers likely have diluted the ^{14}C content of at least some of these samples, and evaluation of these samples requires additional information.

Oxygen-isotope data provide understanding complementing the ^{14}C data. Two groups of low-^{14}C samples are evident in figure 11: those with ^{14}C concentrations of less than 20 PMC and $\delta^{18}O$ values less than (more negative than) -17.6‰ (ranging from -19.2 to -17.7‰), and those with ^{14}C concentrations of less than 20 PMC but $\delta^{18}O$ values greater than -17.5‰ (ranging from -17.4 to -15.8‰).

The group of samples with low ^{14}C concentrations (<20 PMC) that have low $\delta^{18}O$ values (-19.2 to -17.7‰) likely represent Pleistocene groundwater. The depletion in [18]O is consistent with recharge during the Pleistocene (Clark and Fritz, 1997, p. 198–200). The combination of depleted $\delta^{18}O$ values and low ^{14}C concentrations provides independent lines of evidence pointing toward Pleistocene recharge. As a group, these samples also tend to be from the deepest wells compiled for this analysis. An alternative hypothesis for the presence of isotopically light groundwater could be recharge from high elevations. This is a highly unlikely origin for the groundwater samples from sites east of the Columbia River (fig. 10) because these sites are hydrogeologically isolated from high-elevation recharge sites (Bauer and others, 1985; Burns and others, 2011). On the other hand, the low-^{14}C, low-$\delta^{18}O$ groundwater samples from sites north and west of the Columbia River conceivably could have originated in mountainous terrain upgradient and to the west of the sampling sites (fig. 10). Based on stable-isotope data for precipitation in the Yakima River Basin, such isotopically light groundwater likely would have had to originate from sites with average elevations substantially greater than 4,000 ft (Vaccaro and others, 2009). This also is demonstrated in the isotopic character of the Naches River near North Yakima, Wash., which integrates precipitation from a basin with an average elevation of about 4,400 ft (Vaccaro and others, 2009). The Naches River near North Yakima, Wash., has an average $\delta^{18}O$ value of about -14‰ (Vaccaro and others, 2009), substantially heavier than the low-^{14}C, isotopically depleted samples ($\delta^{18}O$ values -19.2 to -17.7‰). However, it would be difficult for high-elevation recharge to migrate relatively unmixed (unmixed with lower-elevation recharge) through the structurally faulted, folded, and compartmentalized geologic materials in this part of the CPRAS (Burns and others, 2011) by way of the spatially complex groundwater-flow regime (Vaccaro and others, 2009) to these deeper wells (but not to other wells). Thus, although it is possible that these isotopically light samples with low-^{14}C concentrations from sites north and west of the Columbia River owe their $\delta^{18}O$ signatures to high-elevation recharge, the more parsimonious interpretation is an origin from Pleistocene recharge.

The group of low-^{14}C samples that are relatively enriched in ^{18}O ($\delta^{18}O$ values greater than -17.5‰) are not so easily interpreted. The range of $\delta^{18}O$ values of these "ambiguous" samples (-17.4 to -15.8‰) is lighter than the full range of $\delta^{18}O$ values from the samples interpreted to represent essentially modern water ($\delta^{18}O$ values ranging from -17.0 to -13.0‰). Thus, it is possible that some of these samples represent isotopically light groundwater that, in combination with low ^{14}C concentrations, could represent Pleistocene groundwater. The partial overlap between the $\delta^{18}O$ values of these "ambiguous" samples and the $\delta^{18}O$ values of the samples interpreted to represent essentially modern water, combined with the lack of quantification of possible carbon mass transfers, however, leaves interpretation of these samples unresolved. These samples cannot be classified without additional data and understanding.

One additional group of samples remains in figure 11: those with intermediate ^{14}C concentrations (20 to 70 PMC) and intermediate $\delta^{18}O$ values (-18.2 to -15.4‰). Some of these samples could represent groundwater with ages intermediate between Pleistocene and the present, and some could represent mixtures of Pleistocene and late Holocene or modern groundwater.

Selected Hydrologic Budget Components

Recharge from Infiltration of Precipitation

Recharge from infiltration of precipitation was estimated for the study area using a polynomial-regression equation based on annual precipitation and the results of recharge modeling. Bauer and Vaccaro (1990) estimated groundwater recharge to the CPRAS for 1956–77 using a deep-percolation model (DPM; Bauer and Vaccaro, 1987). The DPM used precipitation, temperature, streamflow, soils, land-use, and altitude data to compute transpiration, soil evaporation, snow accumulation, snowmelt, sublimation, and evaporation of intercepted moisture. Daily changes in soil moisture, plant interception, and snowpack were computed and accumulated. Deep percolation (water that percolated below the root zone) was assumed to be available for recharge. Bauer and Vaccaro (1990) stated that the model used assumptions regarding certain processes that would yield what they

termed "conservative" recharge estimates, which meant that the estimates might be low relative to actual recharge. The model was most sensitive to precipitation input, but they also found that soil and land-use parameters could be sensitive in some climatic and topographic settings. The great spatial and temporal variation in these settings prevented a rigorous error analysis, but the authors estimated a maximum model uncertainty of 25 percent (Bauer and Vaccaro, 1990).

Bauer and Vaccaro (1990) used the results of the recharge modeling to develop a 2nd order polynomial-regression equation based on precipitation and simulated recharge with a correlation coefficient of 0.92. Using their equation, annual recharge, R, is related to annual precipitation, P:

$$R = (P^2 \times 0.00865) + (P \times 0.1416) - 1.28. \qquad (1)$$

Bauer and Vaccaro (1990) compared recharge estimates using the regression equation with model-estimated recharge from three climatic regimes in the study area and found the regression equation tended to underestimate maximum annual recharge and overestimate minimum recharge, but matched mean annual recharge closely over all climatic regimes.

Annual-recharge distributions required for this study were computed for 23 years (1985–2007) using the regression equation with the Parameter-elevation Regressions on Independent Slopes Model (PRISM) gridded annual-precipitation data from the PRISM Climate Group at Oregon State University (PRISM Climate Group, 2004). The only modification done to the data was resampling the 800-m PRISM data to the 1-km cells of the SOWAT-model grid used to estimate irrigation water use and associated recharge (*see* next section). Where annual precipitation was less than 6.37 in., the estimated recharge from equation 1 is less than zero and is assumed to be equal to zero (Bauer and Vaccaro, 1990).

Temporal trends in precipitation are naturally mirrored by estimated recharge (fig. 12). Mean annual precipitation for the study area during the period 1985–2007 is 16.8 in/yr and mean annual recharge is 4.6 in/yr (14,980 ft³/s). The spatial distribution in recharge also mirrors that of annual precipitation, with the highest recharge (more than 20 in/yr) occurring in the upper Yakima River Basin, the Blue Mountains southeast of Walla Walla, and adjacent to the Columbia River where it flows west from the study area through the gorge in the Cascade Range (fig. 13). Mean annual recharge from infiltration of precipitation is less than 1 in/yr over a large part of the study area adjacent to the Columbia and Yakima Rivers where precipitation is limited to less than 10 in/yr (fig. 13). These areas lie in the lowest part of the study area and are the locations of much of the irrigated agriculture in the Columbia Plateau.

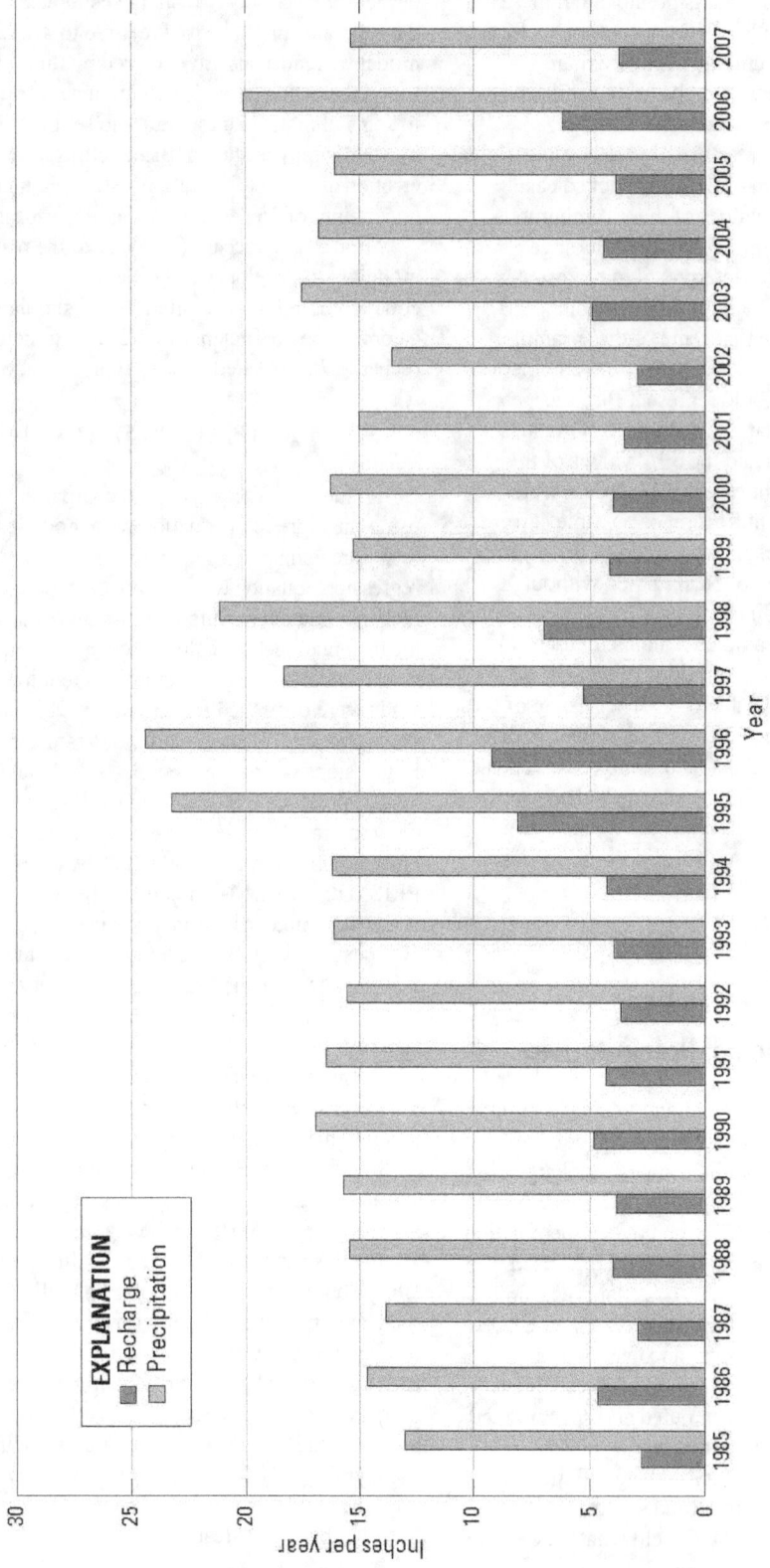

Figure 12.   Annual precipitation and infiltration recharge in the Columbia Plateau Regional Aquifer System, 1985–2007.

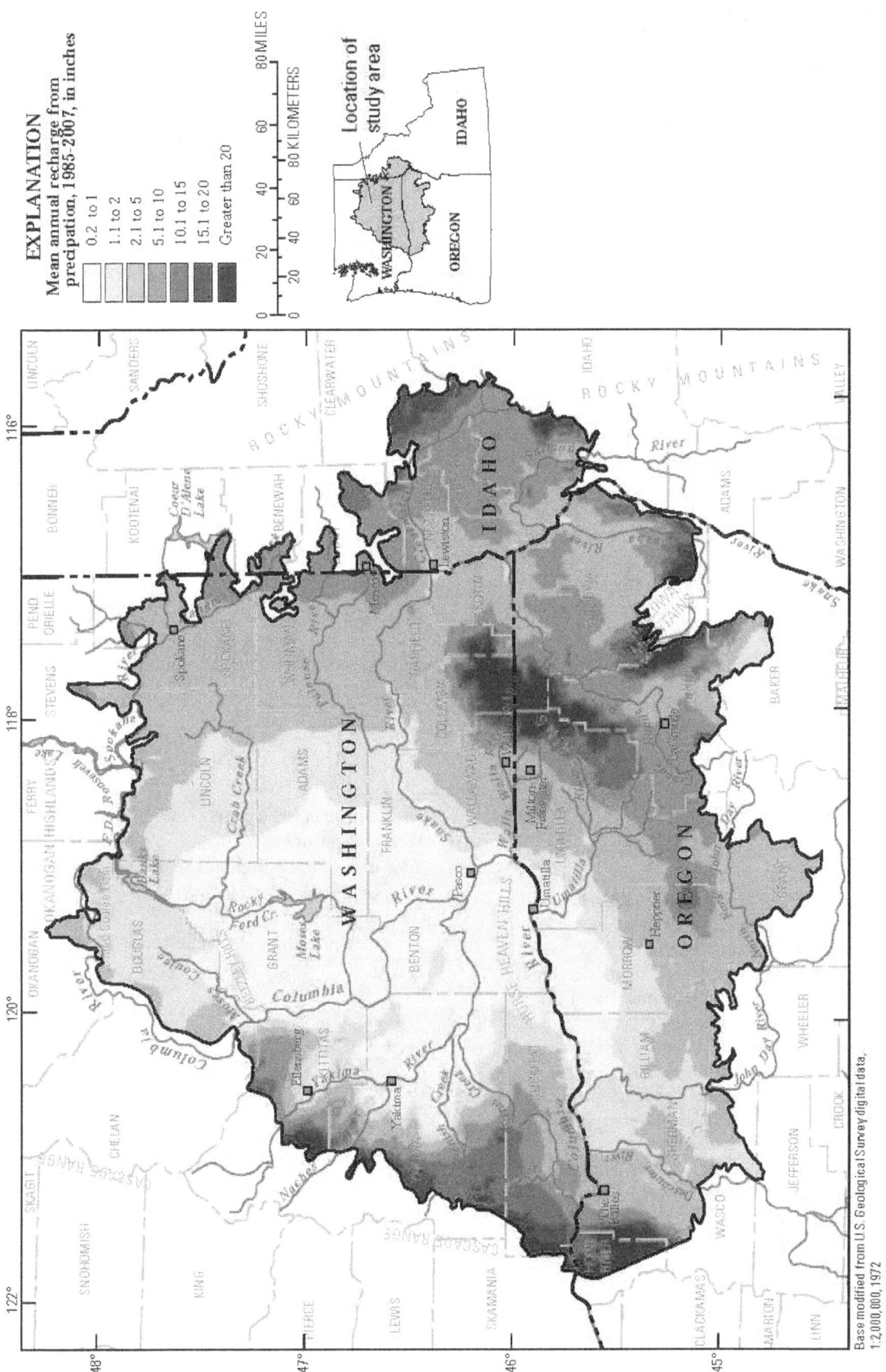

Base modified from U.S. Geological Survey digital data,
1:2,000,000, 1972

Figure 13.   Distribution of mean annual recharge from infiltration of precipitation in the Columbia Plateau Regional Aquifer System, 1985–2007.

Irrigation Water Use and Associated Recharge

Irrigation is the largest water-use category in the CPRAS. The largest groundwater-pumping centers are located where groundwater is the principal source of irrigation, and the areas of greatest recharge from irrigation-return flow occur where surface water is the primary source of irrigation. Each of these budget components, pumping and recharge, has implications for groundwater availability both regionally and locally.

Estimates of groundwater pumpage and surface-water diversions for irrigation and recharge associated with irrigation were made for the period of interest (1985–2007) using a spatially distributed SOil WATer (SOWAT) model developed for this study. The SOWAT model uses simple relations among climatic, soils, land-cover, and irrigation data to compute monthly irrigation requirements and surplus moisture available for recharge. Estimates of groundwater pumping for irrigation and recharge from irrigation return flow from this application of the SOWAT model will be used as initial input to a regional simulation model of the groundwater-flow system.

Landscape evapotranspiration (ET) is a key component of the hydrologic water balance. Spatial distribution of landscape ET can be used as an indicator of vegetation performance in terms of biomass accumulation, which is directly associated with water use (Senay, 2008). Furthermore, ET can be used to estimate the spatio-temporal dynamics of the rates and total amounts of groundwater recharge and withdrawal from aquifer systems in irrigated areas.

The SOWAT model was specifically developed to make use of estimates of actual evapotranspiration (ETa) available from a new Simplified Surface Energy Balance (SSEB) method, which utilizes remotely sensed land-surface temperature data (Senay and others, 2007). Most water-balance models compute potential ET (ETp) using climate data with an empirical equation (Blaney and Criddle, 1950; Thornthwaite and Mather, 1955; Hamon, 1961; Bauer and Vaccaro, 1987), and actual ET is computed as a function of ETp and soil moisture (Thornthwaite and Mather, 1955; Bauer and Vaccaro, 1987; Ellis and others, 2008; McCabe and Markstrom, 2007). Senay and others (2007) have shown that estimates of ETa from the SSEB method compare well with ground-based ET-flux measurements from lysimeters (Gowda and others, 2009) and with results from other satellite-based energy-balance methods (Allen and others, 2007). A linear regression between SSEB-estimated ETa and measured ETa yielded a correlation coefficient (r^2) of 0.84 and a root mean square error (RMSE) of 1.2 mm/d, which was 22 percent of the observed mean. The RMSE was most influenced by overprediction in the early part of the growing season.

The principle advantage of the SSEB method to estimate ETa for a regional water-balance model is the reliance on remotely sensed temperature, a first-order indicator of the energy consumed by the process of evapotranspiration, rather than the empirical relation between ETp and soil moisture used in previous approaches. This reduces the uncertainty in the ETa component of the water budget used in SOWAT.

The components of the soil-water balance represented in the SOWAT model, its implementation, and data requirements, are described in the following sections. The results of the application of the model to the CPRAS are then described.

Monthly Soil-Water Balance Model

Monthly water-balance models are a useful means of examining components of the hydrologic cycle (Stonestrom and Harrill, 2007). Such models have been used at a wide range of scales and for a variety of purposes, including to estimate the global water balance (Mather, 1969; Legates and McCabe, 2005), soil-moisture storage (Alley, 1984), runoff (Alley, 1984, 1985; Wolock and McCabe, 1999), and irrigation demand (McCabe and Wolock, 1999). Similar models also have been used to evaluate the hydrologic effects of climate change (Yates, 1996; Strzepek and Yates, 1997; Wolock and McCabe, 1999).

The soil-water balance model used in this study (fig. 14) includes the concepts of climatic water supply (precipitation) and climatic water demand (evapotranspiration), seasonality in climatic water supply and demand, soil-moisture storage, and irrigation practices. The primary purpose of the model in this application is to estimate (1) the irrigation demand when climatic water demand exceeds climatic water supply and available soil moisture, and (2) surplus moisture available for deep percolation and recharge to the groundwater system. The mass-balance equation solved by the SOWAT model is:

$$\Delta SM = PR + IR - ETa - DR - GF, \qquad (2)$$

where

ΔSM is change in soil moisture storage (L),

$\quad PR$ is precipitation (L),

$\quad IR$ is irrigation application (L),

ETa is actual evapotranspiration (L),

$\quad DR$ is direct runoff (L), and

$\quad GF$ is flux below the modeled soil zone (L).

Implementation and Data

Climate inputs to the model are PR and ETa. DR is the runoff resulting from infiltration excess overflow and is estimated as a fraction ($DRfrac$) of precipitation,

$$DR = DRfrac \times PR.$$

The net infiltration (NI) is the residual,

$$NI = PR - DR.$$

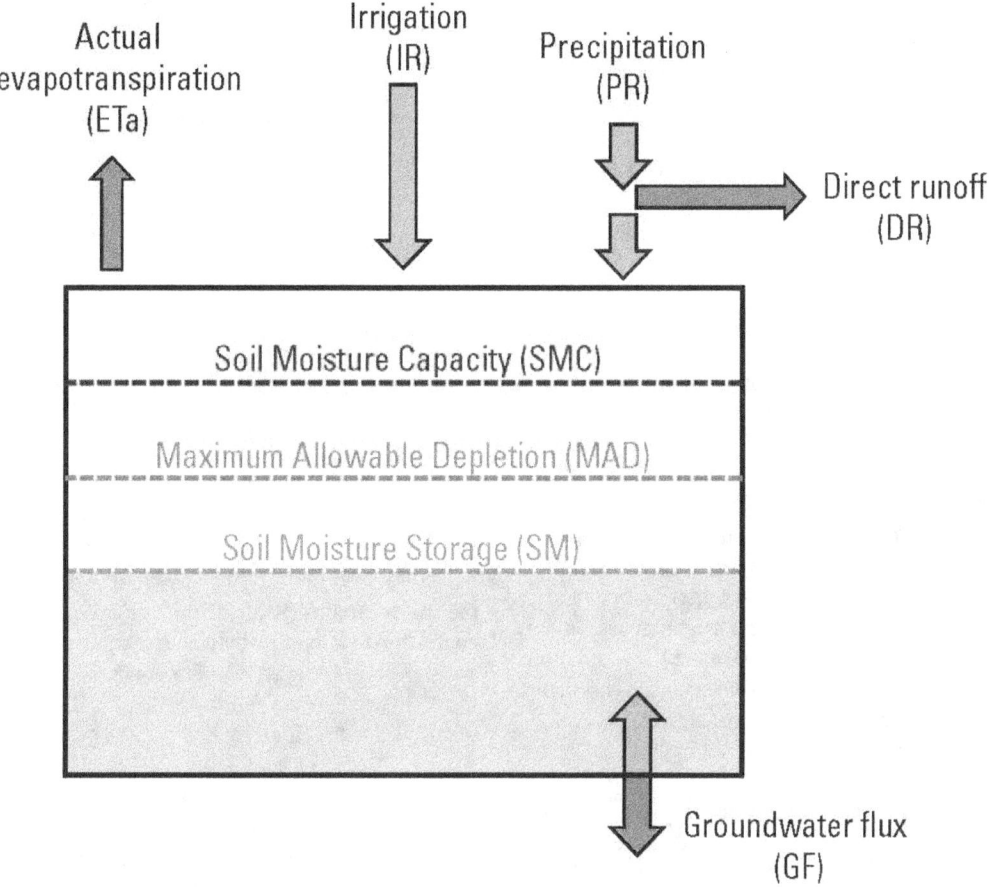

Figure 14. Components of the soil-water balance represented in the model.

Net infiltration is added to the soil moisture (*SM*) from the previous month (*SM'*), and *ETa* is subtracted to determine current month *SM*,

$$SM = SM' + NI - ETa.$$

During the irrigation season, current *SM* is compared with the maximum allowable depletion (*MAD*) to determine the irrigation requirement. The *MAD* is the fraction of the soil-water capacity that can be removed from soil-moisture storage before crop productivity will suffer (Bauder and Carlson, 2010). If *SM* is less than the irrigation target (*IT*),

$$IT = SMC \times MAD,$$

then irrigation is supplied to fill the soil profile to capacity (*SMC*). Irrigation-system efficiency (*IE*) is the fraction of irrigation water diverted from surface-water sources or pumped from groundwater sources that is actually used by plants as evapotranspiration. Applied irrigation is computed as,

$$IR = (SMC - SM) / IE.$$

Excess irrigation water (owing to inefficiency) is diverted to *GF*,

$$GF = SM + IR - SMC.$$

If soil moisture is at or above the irrigation target, no irrigation is applied during the current month. If *SM* is greater than the *SMC*, excess moisture is assumed available for recharge and added to *GF*. In months outside the irrigation season, current month *SM* is computed and if *SMC* is exceeded, excess again is assumed available for recharge and added to *GF*. If *ETa* exceeds net infiltration and soil-moisture storage in non-irrigation months, groundwater flux (discharge) is assumed to account for the soil-moisture deficit.

The study area was divided into 1-km (3,280 ft) square cells, and the model was used to compute the monthly soil-water balance in each cell for the period January 1985–December 2007. The model is written in Python and reads all spatial data directly from raster files (ESRI-grid format) (L.L. Orzol, U.S. Geological Survey, written commun., April 2010). Data required to describe land-cover and soil properties, spatio-temporal dynamics of climate, and irrigation practices are described in the following sections.

Land-Cover and Soil Properties

Three land-cover classes are recognized by the SOWAT model: irrigated agriculture, non-irrigated agriculture and native vegetation, and developed lands and water bodies. The primary purpose of the model is to compute irrigation and recharge in irrigated areas, but the water balance also is computed in native/non-irrigated and developed/water cells. The model extent comprises 114,112 cells (square kilometers) or about 44,000 mi^2 (fig. 15). Irrigated cells were identified using four sources of information: (1) the computed ratio of seasonal ETa to precipitation in each cell, (2) aerial photography, (3) water-rights data, and (4) crop mapping (U.S. Department of Agriculture, 2007b, 2007c, 2007d). In 2007, there were 19,091 cells identified in which part of the cell area included irrigated agriculture. The developed parts of the study area within Idaho (Moscow and Lewiston) include very little irrigated agriculture, and none of these cells in the model were identified as irrigated from the available data.

The land-cover grid is used by SOWAT only to identify cells where irrigation is possible; no irrigation is computed if there is not a moisture deficit during the irrigation season. The 2007 land-cover grid was used for all years of the simulation (1985–2007) to identify potentially irrigated cells. This could result in errors only where previously irrigated land was converted to another use, a rare occurrence in the study area since 1985.

The soil-moisture (or available water) capacity (SMC) of the soil is defined as the volume of water that should be available to plants if the soil were at field capacity. SMC (as depth in millimeters) of the upper 150 cm of soil derived from the U.S. General Soils Map (STATSGO2) (Miller and White, 1998) was used to define storage available in the study area (fig. 16). Water-capacity data were available for the upper 100, 150, and 250 cm of soil from Miller and White (1998). The 150 cm (59 in.) depth was chosen to best represent the range of crop-root depths encountered in the study area. Initial soil-moisture content (in millimeters) for the month prior to the start of the simulation period (December 1984) was estimated by running the model for 1985–2007 and computing the mean December soil-moisture values for 2000–07.

The ETa was specified for each cell using estimates from the SSEB. SSEB makes the assumption that latent heat flux (ETa) varies linearly with near-surface temperature differences (Senay and others, 2007). USGS Hydro1K Digital Elevation Model (DEM) data were used for surface-temperature correction to remove elevation-induced surface-temperature changes. SSEB uses the elevation corrected 1-km land-surface temperature data from the Moderate Resolution Imaging Spectroradiometer ($MODIS$) sensor to identify "cold pixels" in heavily irrigated areas where ETa is equal to ETr (reference ET) and "hot pixels" in dry barren or fallow areas where ETa is near zero. The ratio of the temperature of each pixel to the cold-pixel temperature is used to compute ETa as a fraction of ETr. ETr is computed using weather data from the Global Data Assimilation System (GDAS) as described in Senay and others (2008).

Senay and others (2009) applied the SSEB method to estimate ETa for the CPRAS. The average ETa for 8 years (2000–07) in 13 Yakima irrigation subbasins was compared to reported irrigation-application estimates by (Vaccaro and Olsen, 2007). Estimates of mean annual ETa from SSEB in the 13 subbasins were well correlated with the independent estimates of irrigation; r^2 from a linear regression was equal to 0.92. Irrigation estimates by Vaccaro and Olsen (2007) were, however, about three times the estimated ETa, indicating that irrigation-system efficiencies in the subbasins were in the range of 30 to 40 percent.

Prior to deployment of the $MODIS$ sensor in 2000, 1-km land-surface temperature (LST) data were collected by the Advanced Very High Resolution Radiometer (AVHRR) sensor. Unfortunately, the LST data for the study area from the AVHRR sensor for 1989–99 were of poor quality during the cooler months (September–June) when there was less contrast in temperature between irrigated and non-irrigated areas. July and August temperature data from AVHRR and the resulting estimates of ETa from the SSEB model generally were good within the irrigated areas. The estimates of July and August ETa from $MODIS$ and AVHRR data for the 2000–07 period were compared and, although there is a consistently high bias in AVHRR-derived ETa estimates, there also is good correlation. This correlation for the 2000–07 period was used as the basis for adjusting AVHRR-based estimates of ETa for the 1989–99 period. The procedure for the adjustments is described below.

Linear regressions were run on July and August ETa values from the 19,091 cells with irrigation for 2000–07 where $MODIS$ ETa was the dependent variable and AVHRR ETa was the independent variable. Linear-regression equations (r-square = 0.78) were developed to estimate July and August ETa during 1989–99:

$$MODIS\ ETjuly = AVHRR\ ETjuly \times 1.0284 - 27,205$$
$$MODIS\ ETaugust = AVHRR\ ETaugust \times 1.1077 - 24.575.$$

AVHRR ETa data quality for July was good in all years except 1994, 1995, and 2002, in which August ETa data were used. Regression also showed correlation (r-square = 0.86) between July and August $MODIS$ ETa and annual $MODIS$ ETa for 2000–07, which allowed the annual $MODIS$ ETa to be estimated for 1989–99 using:

$$MODIS\ ETannual = MODIS\ ETjuly \times 3.457522 + 49.1694$$
$$MODIS\ ETannual = MODIS\ ETaugust \times 3.838676 + 78.2317.$$

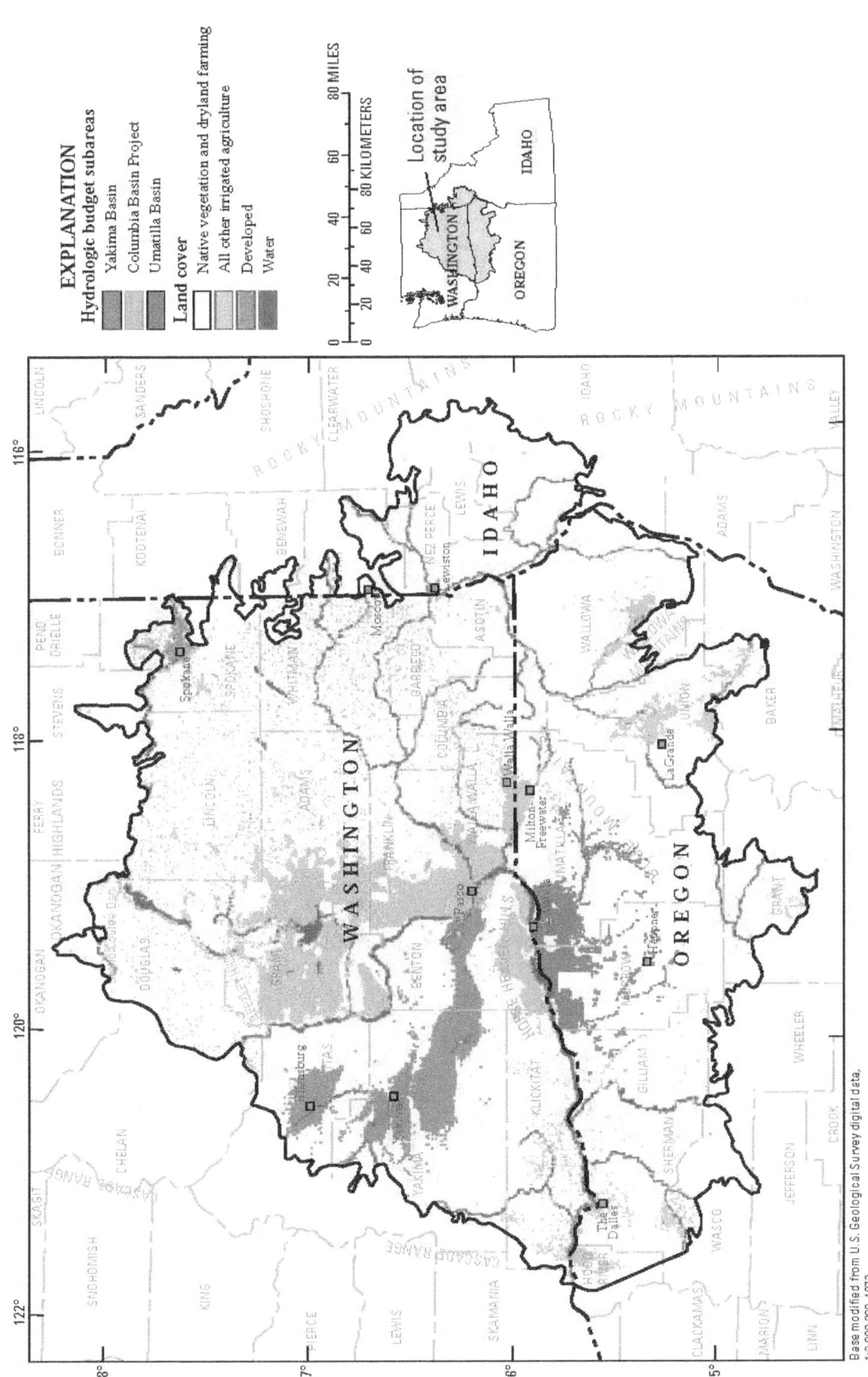

Base modified from U.S. Geological Survey digital data.
1:2,000,000, 1972

Figure 15. Simplified land cover and the location of hydrologic budget subareas in the Columbia Plateau Regional Aquifer System.

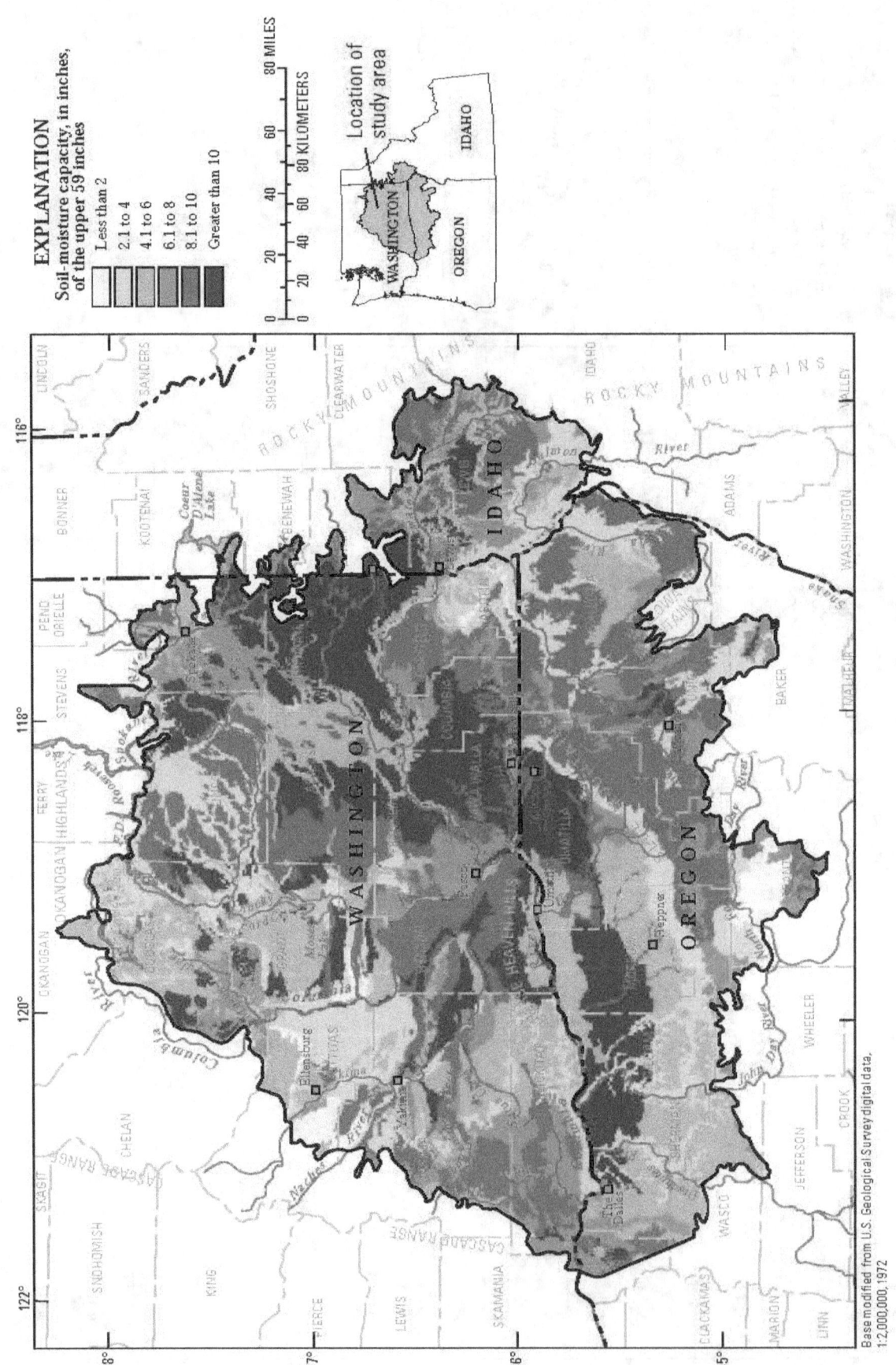

Base modified from U.S. Geological Survey digital data.
1:2,000,000, 1972

Figure 16.　Soil-moisture capacity in the Columbia Plateau Regional Aquifer System.

Monthly values of *MODIS ETa* were estimated for 1989–99 using the monthly mean fractions of annual *ETa* computed for 2000–07 (fig. 17). For example, if the total annual *MODIS ETa* in 1992 at a given cell was 40 in., the June 1992 monthly *ETa* at that cell was estimated to be 20 percent (fig. 17) of the total, or 8 in. The mean standard deviation of the monthly fractions of annual *ETa* for 2000–07 (fig. 17) was 0.01 (1 percent) indicating that relatively little error was introduced by using the mean values for 1989–99. Finally, AVHRR data were not available for 1985–88, so as a first approximation, *ETa* for these years was assumed the same as 1989. The resulting regression-based adjustments made to AVHRR-derived *ETa* show that the magnitude and seasonal and inter-annual variability of the adjusted *ETa* compares well with *MODIS*-derived *ETa* for 2000–07 (fig. 18).

Precipitation data were derived from PRISM 4-km (2.5 arc-sec) gridded monthly data (PRISM Climate Group, 2004). The only modification done to the data was resampling down to the 1-km cells of the SOWAT-model grid.

The fraction of monthly precipitation that becomes direct runoff, *DRfrac,* is a simplistic abstraction to account for water from precipitation that is not available for infiltration to the soil zone. Based on results from daily soil-water balance modeling (Bauer and Vaccaro, 1990), direct runoff is a relatively small percentage of precipitation (0–8 percent) in irrigated parts of the study area. Precipitation typically ranges from 6 to 10 in/yr in these areas, which would yield less than 1 in/yr of runoff. Compared with irrigation, which can exceed 20–30 in/yr, runoff is a very small part of the overall water budget. A value of 7 percent was used in the model for all cells throughout the simulation period.

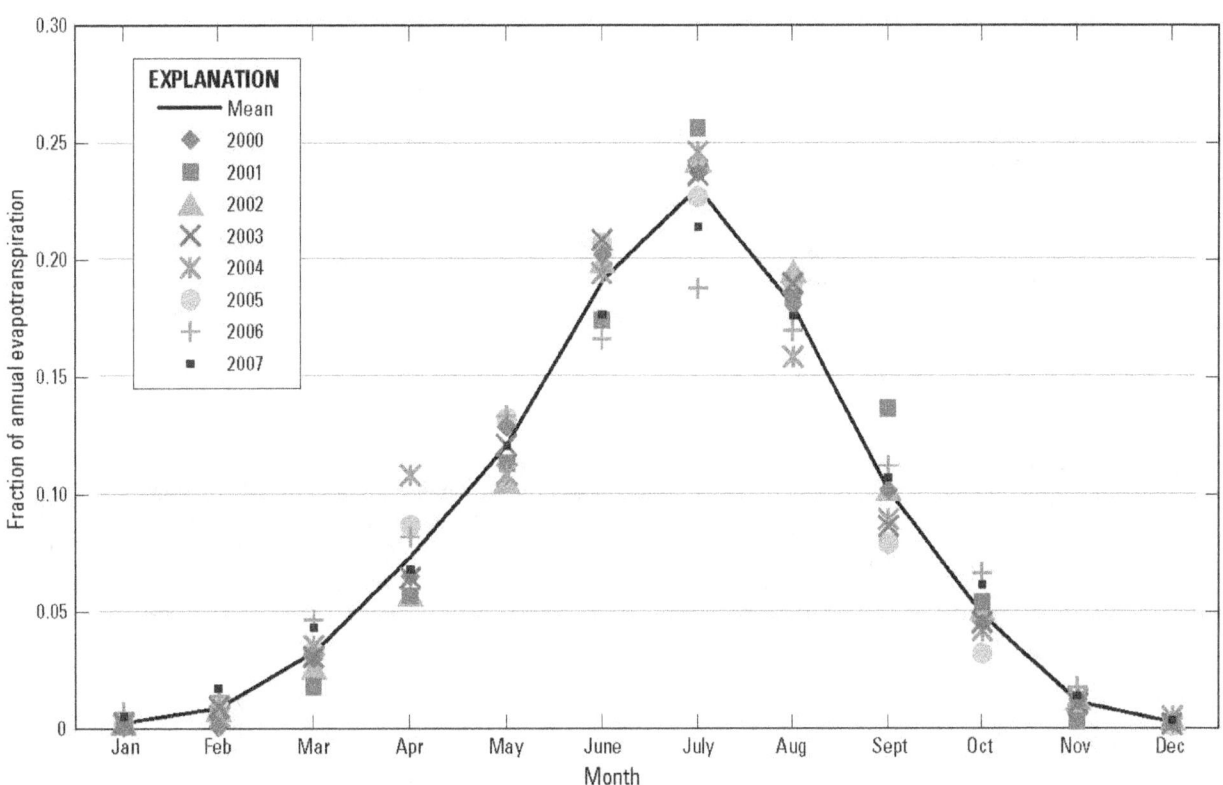

Figure 17.   Mean monthly fraction of annual actual evapotranspiration (2000–2007) used to estimate 1985–99 monthly values.

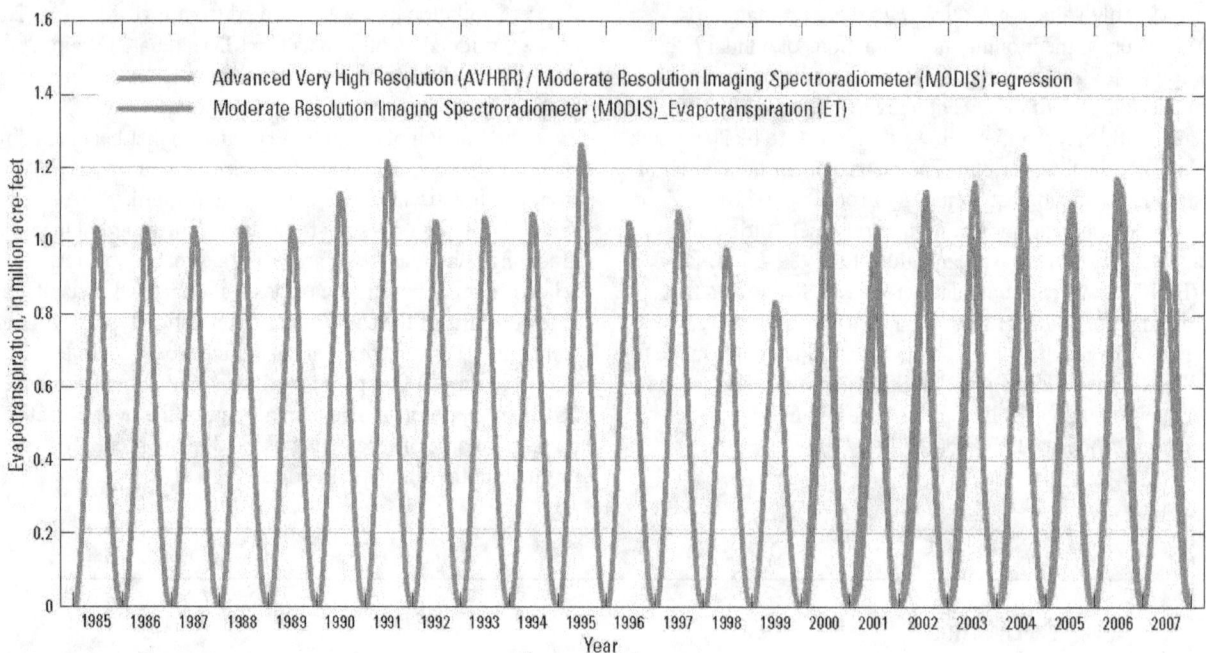

Figure 18.   Comparison of regression-based estimates of actual evapotranspiration (*ETa*) with Moderate Resolution Imaging Spectroradiometer (*MODIS*)-derived *ETa* values.

Irrigation

The annual irrigation season was defined for the entire model domain and simulation period by specifying the beginning and ending months when irrigation water is available. For the majority of the study area in Oregon and Washington, irrigation-water rights limit the irrigation season to April 1–October 31.

Irrigation-system efficiency is defined in SOWAT as the fraction of water diverted from surface water or pumped from groundwater that is consumptively used by crops and accounts for losses that occur both on-farm and, for surface-water supplied irrigation, within the conveyance system.

For example, if a crop has a consumptive-use requirement of 24 in/yr and the overall efficiency of the irrigation system supplying the crop is 0.50 (50 percent), then 48 in/yr (24 in./0.50) would have to be diverted or pumped to supply crop requirements. The water not consumptively used by the crop may percolate to the groundwater system and become recharge, become runoff, or evaporate (U.S. Department of Agriculture, 1997). Crop water use and evaporative losses are accounted for by the SSEB-estimated *ETa*, while losses to percolation (recharge) are estimated as the residual of the water budget in the SOWAT model. Runoff generated from irrigation generally is small and if it occurs, often percolates to the groundwater system before leaving the irrigated area.

On-farm losses are dependent on the method of application, which often is closely related to the source of irrigation water (groundwater or surface water). Surface irrigation, in which water is applied and distributed over the soil surface by gravity, is the least-efficient application method and efficiencies range from 50 to 60 percent. Pressurized (sprinkler) irrigation systems have higher application efficiencies, 60 to 90 percent, owing largely to reduced losses to deep percolation (U.S. Department of Agriculture, 1997). Surface-water supplied irrigation also has associated conveyance losses owing to canal and lateral leakage and operational spills; these can range from 25 to more than 70 percent depending on system design, age, geology, and other factors (Gannett and others, 2001; Montgomery Water Group, 2003). Surface irrigation typically has been used in areas supplied with surface water, although in the past 20 years there has been significant conversion to sprinkler irrigation. Sprinkler irrigation has been the norm in groundwater-supplied areas from their inception, and in most cases, there are no conveyance losses because the well is located in the field that is being irrigated.

SOWAT requires that overall irrigation-system efficiencies, accounting for both on-farm and conveyance-system losses, be specified for both groundwater and surface-water-supplied lands for each year of the simulation. For this analysis, groundwater-supplied lands were assigned an overall system efficiency of 75 percent, which is a median value for sprinkler application. Overall irrigation-system efficiency for surface-water-supplied lands was estimated based on average irrigation project-wide conveyance losses of 26 percent reported for the Columbia Basin Project (Montgomery Water Group, 2003) and on-farm application efficiency. Application efficiency for surface-water-supplied lands was estimated to increase from 66 to 71 percent from 1985 to 2007 as more systems were converted from surface (gravity) to sprinkler irrigation (Montgomery Water Group, 2003). Overall system efficiency for surface-water-supplied lands was computed by first calculating the water needed at the farm headgate (crop requirement/application efficiency), then calculating the diversion needed to supply the headgate delivery (headgate requirement/conveyance system efficiency), and finally computing overall efficiency (crop requirement/diversion). The resulting overall system efficiency for surface-water-supplied lands increased from 49 to 53 percent during

1985–2007 owing to increases in application efficiency during the period. If both groundwater- and surface-water-supplied lands occur within a single cell, SOWAT computes the area-weighted mean irrigation efficiency using the groundwater-supplied irrigation fraction (GW) described in the next section.

The source of irrigation water to each model cell is specified for each year of the simulation (fig. 19). The value is equal to the fraction of irrigation supplied from groundwater and may range from 0.0 (100-percent surface-water supplied) to 1.0 (100-percent groundwater supplied). The groundwater fraction (GW) is used to compute the weighted mean irrigation efficiency for each cell (IE) and to partition monthly irrigation (IR) into its groundwater (IG) and surface-water (IS) components. The initial groundwater-fraction dataset for the study area was developed from water-rights data and maps of irrigation-district boundaries and was modified after comparison of modeled irrigation with independent estimates. The GW dataset for the Oregon part of the study area was derived from analysis of water-right acreages by quarter-quarter section (40 acre) (K. Wozniak, Oregon Water Resources Department, written commun., October 2009). The Washington part was derived using data from the Washington Department of Ecology Water Rights Tracking System (WRTS) supplemented by irrigation-district maps and analysis of aerial photos. Many surface-water irrigators do not receive a full allocation during drought years and have supplemental or standby groundwater supplies. These areas were identified using water-rights data and in selected dry or drought years, the groundwater fraction was increased to reflect the use of these supplemental sources. During 1985–2007, the years in which supplemental groundwater sources were most heavily used in the study were 1987, 1988, 1992–94, 2001, and 2005.

MAD is the desired soil-water deficit at the time of irrigation and is expressed as the percentage of soil-moisture capacity (*SMC*). Providing irrigation water before the deficit exceeds MAD minimizes plant-water stresses that could reduce yield and quality (U.S. Department of Agriculture, 1997). Values of MAD vary by crop, growth stage, and soil texture. For example, USDA guidelines for MAD range from 35 percent for potatoes to 50 percent for orchards, alfalfa, and sugar beets. SOWAT uses one value of MAD for all months of the simulation, and a value of 50 percent was specified based on the overall mix of crops in the study area.

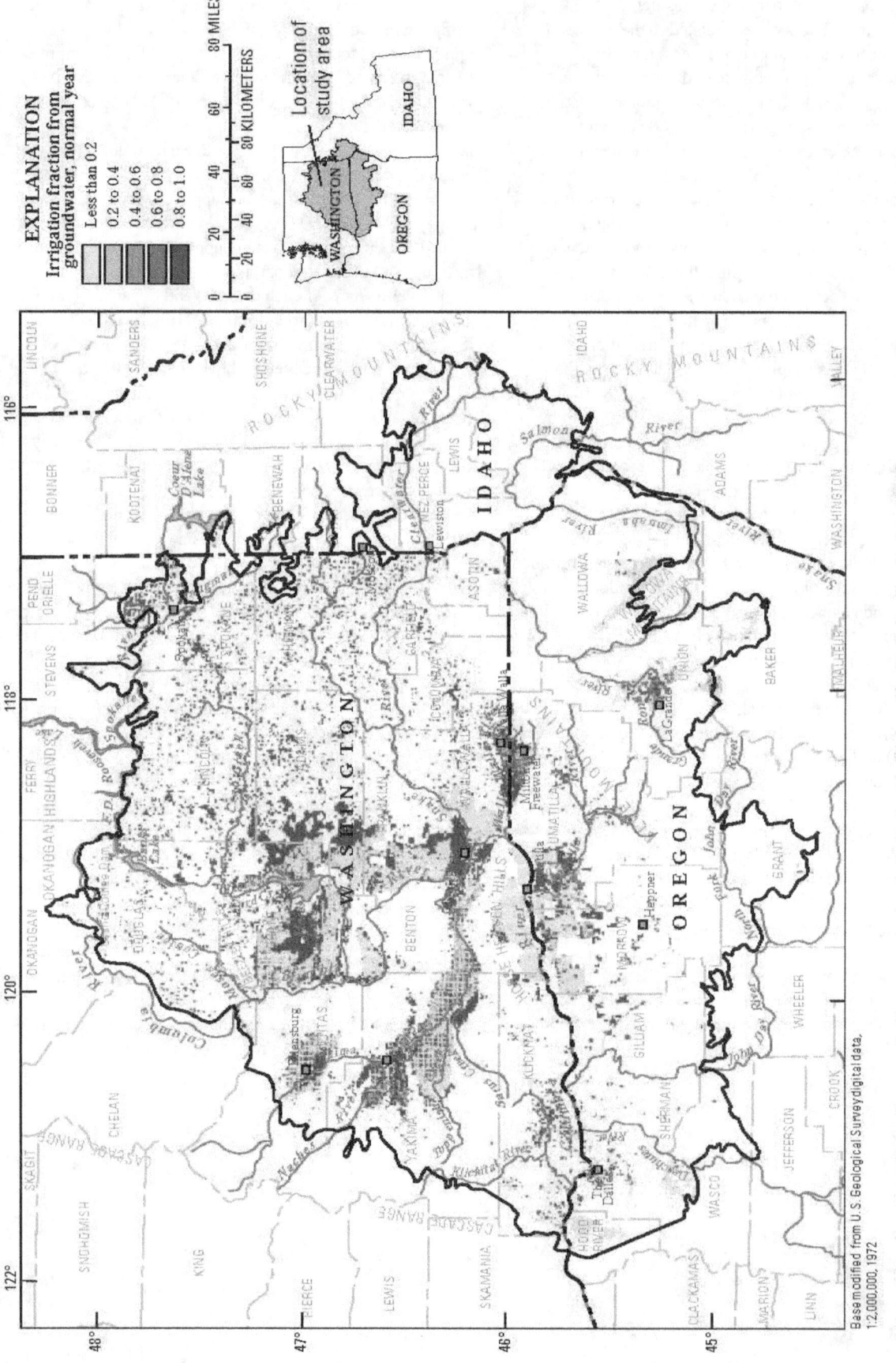

Base modified from U.S. Geological Survey digital data.
1:2,000,000, 1972

Figure 19. Fraction of irrigation supplied by groundwater in the Columbia Plateau Regional Aquifer System, 2007.

Results

The mean monthly soil-water balance in irrigated lands is dominated by the climatic water demand of ET and the application of irrigation water to satisfy crop-water requirements (fig. 20). Precipitation replenishes soil moisture and contributes to groundwater recharge in the late fall and winter months (November–January). Some early spring ET demand is supplied by *SM*, but is replenished with the beginning of the irrigation season in April, as SOWAT applies irrigation at rates sufficient to maintain *SM* above MAD. Groundwater recharge in the irrigated lands peaks again in the summer with ET demand and irrigation applications, and tapers in the fall before precipitation increases. Mean monthly irrigation throughout the study area peaks in July at 1.6 MAF (1985–2007 average), of which 0.45 and 1.15 MAF are from groundwater and surface-water sources, respectively. Direct runoff, specified as 7 percent of precipitation, is a minor part of the hydrologic budget in the irrigated lands.

Irrigation Water Use

Annual irrigation water use in the study area averaged 5.3 MAF during 1985–2007, with 1.4 MAF (or 26 percent) supplied from groundwater and 3.9 MAF supplied from surface water (table 6, fig. 21*A*). There is no apparent

long-term trend in either groundwater or surface-water irrigation. Dry years, when supplemental groundwater rights are exercised, are apparent (1987, 1988, 1992–94, 2001, and 2005); however, there are other years (1997, 2000, and 2007) that were not considered dry, but groundwater was a larger percentage of total irrigation (fig. 21*A*). These are years when areas predominately supplied by groundwater had above normal *ETa*, which were in effect, localized droughts.

Table 6. Estimated mean annual irrigation water use for the Columbia Plateau Regional Aquifer System, 1985–2007.

[Groundwater use in millions of acre-feet (MAF)]

Budget area	All irrigation	Surface-water supplied		Groundwater supplied	
	MAF	MAF	Percent	MAF	Percent
Umatilla Basin	0.46	0.33	72	0.13	28
Yakima Basin	1.18	.93	79	.25	21
Columbia Basin Irrigation Project	1.74	1.71	98	.03	2
All other irrigated lands	1.92	.93	48	.99	52
Total	5.3	3.9	74	1.4	26

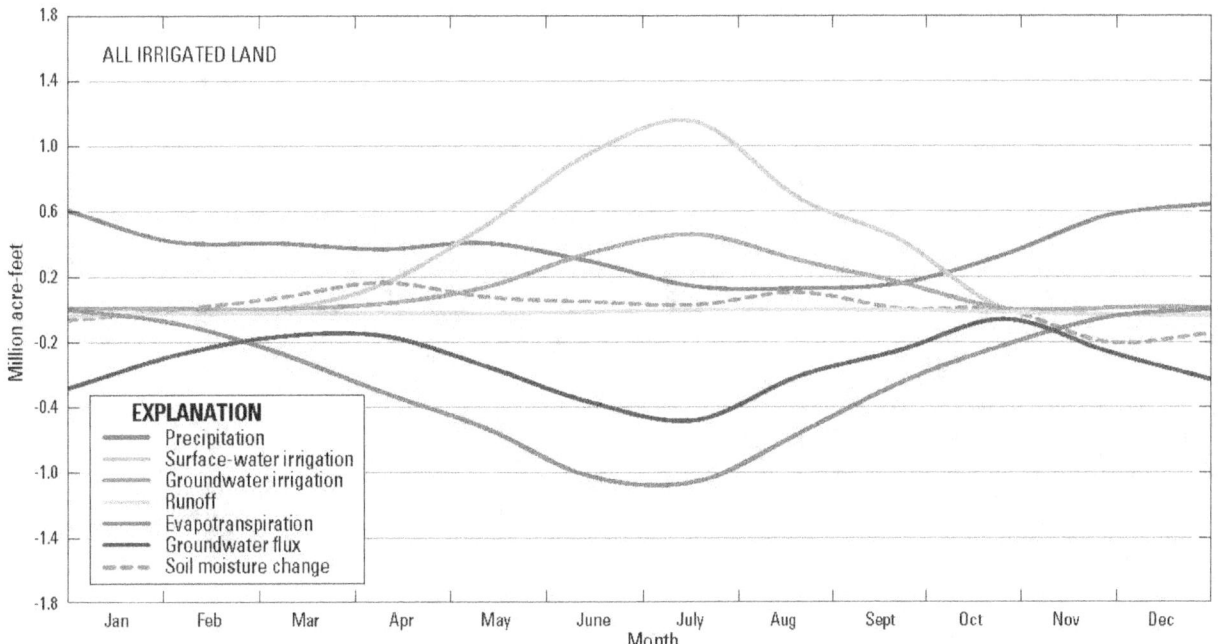

Figure 20. Mean monthly components of the monthly soil-water balance in irrigated areas of the Columbia Plateau Regional Aquifer System, 1985–2007.

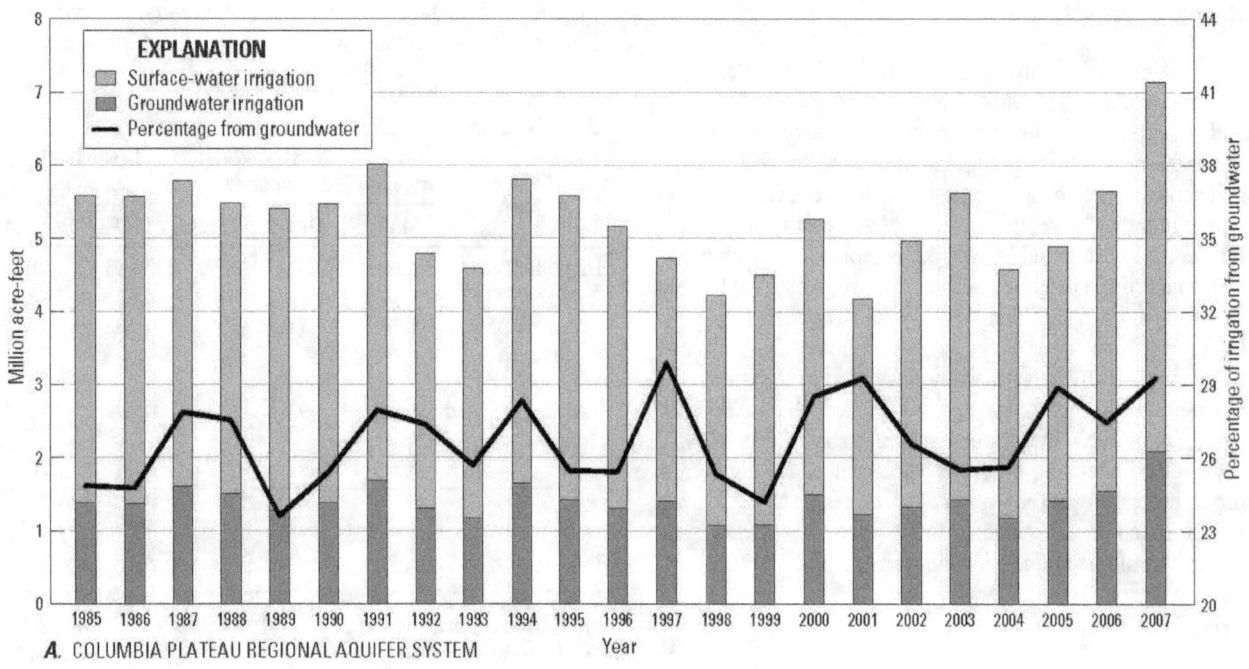

A. COLUMBIA PLATEAU REGIONAL AQUIFER SYSTEM

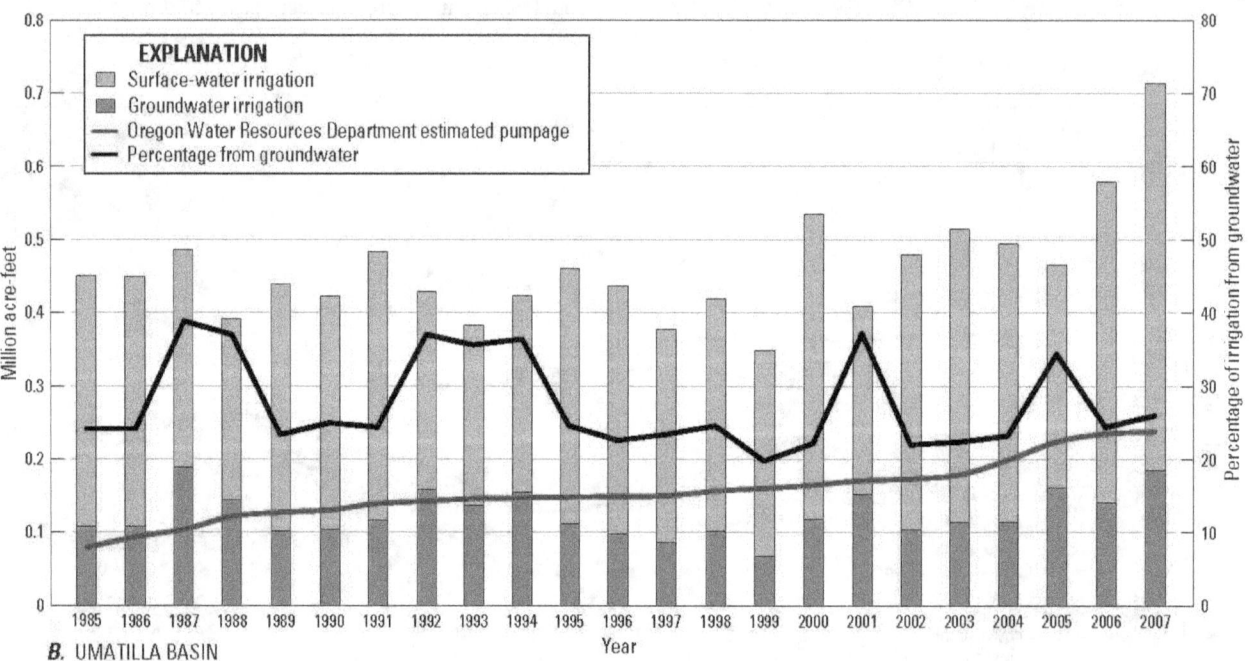

B. UMATILLA BASIN

Figure 21.   Estimated groundwater and surface-water irrigation, 1985–2007. (*A*) Columbia Plateau Regional Aquifer System. (*B*) Umatilla Basin. (*C*) Columbia Basin Project. (*D*), Yakima Basin.

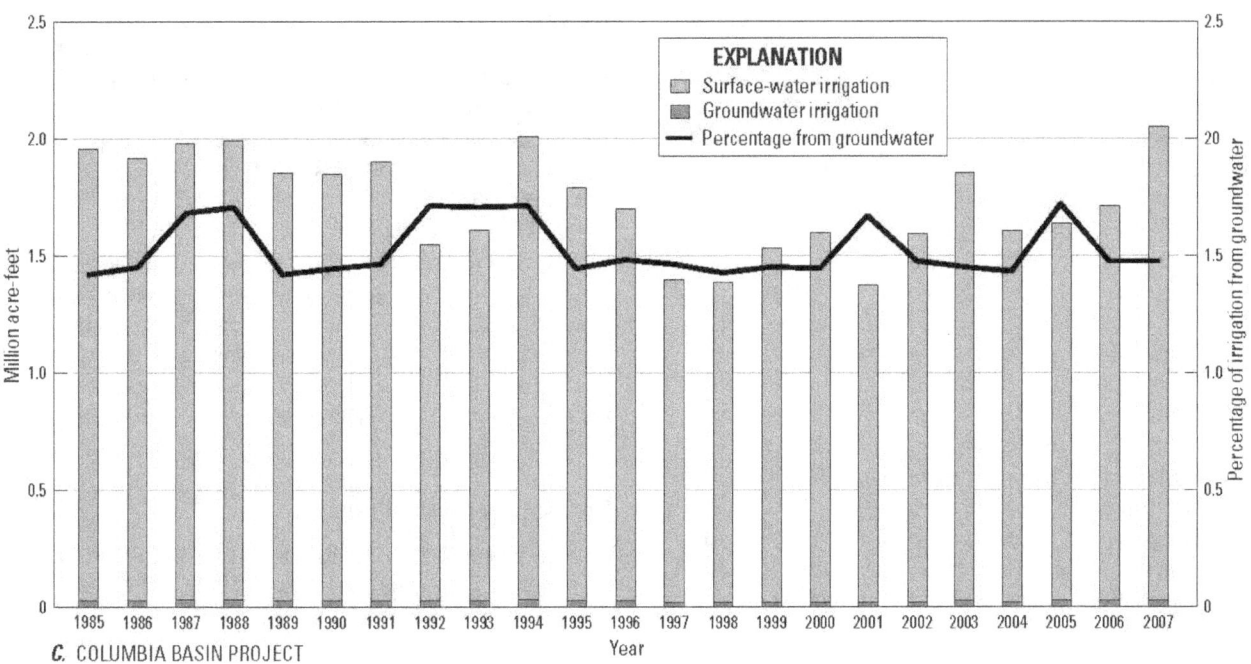

C. COLUMBIA BASIN PROJECT

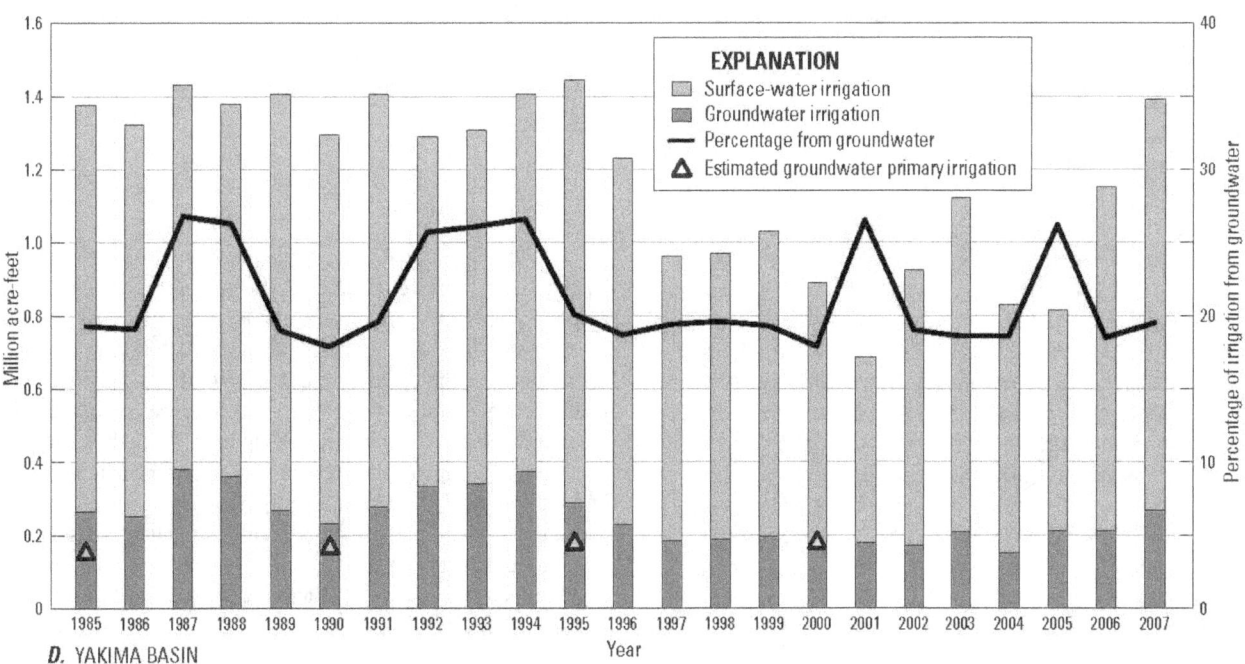

D. YAKIMA BASIN

Figure 21.—Continued

To allow comparison with estimates from other studies, irrigation water use was summarized for selected hydrologic budget subareas that include the Umatilla Basin, Yakima Basin, and Columbia Basin Project (CBP) (fig. 15).

Annual irrigation water use in the Umatilla Basin (fig. 15) averaged 0.46 MAF during 1985–2007, with 0.13 MAF (or 28 percent) supplied from groundwater and 0.33 MAF supplied from surface water (fig. 21B). Based on water-rights data for 2007, there were approximately 109,000 acres supplied primarily by groundwater and 169,000 acres supplied primarily by surface water (K. Wozniak, Oregon Water Resources Department, written commun., October 2009). Using these acreages with estimated irrigation use of 0.53 and 0.19 MAF for surface water and groundwater, respectively, in 2007 (fig. 21B), the average application rates for groundwater- and surface-water supplied areas were 21 and 38 in/yr, respectively. In normal years, groundwater supplies about 25 percent of irrigation in the basin, but in dry years, use of supplemental wells causes the percentage to climb to 35–38 percent (fig. 21B). OWRD made a preliminary estimate of groundwater pumping in the basin for 1985–2007 using estimates of irrigated acreage compiled from water-rights data and an estimated application rate of 24 in/yr for all irrigated lands (K. Wozniak, Oregon Water Resources Department, written commun., October 2009) (fig. 21B). SOWAT pumpage estimates generally are less than the independent OWRD estimates (fig. 21B) and do not show the long-term increasing trend contained in the OWRD estimates; however, there is good agreement overall, which is significant because the approaches used were vastly different.

Irrigation water in the CBP (fig. 15) is almost entirely supplied from water diverted from the Columbia River at Lake Roosevelt. Irrigation water use computed by the SOWAT model averaged 1.74 MAF during 1985–2007, of which 1.71 MAF (98 percent) was supplied from surface water (fig. 21C). Estimated net diversions to CBP averaged 2.4 MAF during 1992–2002 (Montgomery Water Group, 2003), which is 29 percent greater than SOWAT predictions. The discrepancy could be caused by underestimation of ETa, overestimation of system irrigation efficiency, or uncertainty in the estimated net diversions to the CBP subarea.

Annual irrigation water use in the Yakima Basin (fig. 15) averaged 1.18 MAF during 1985–2007, with 0.25 MAF (or 21 percent) supplied from groundwater and 0.93 MAF supplied from surface water (fig. 21D). In normal years, groundwater supplies about 19 percent of irrigation in the basin, but in dry years use of supplemental wells causes the percentage to increase to 26 or 27 percent. Estimates of groundwater pumpage for irrigation in the Yakima Basin were made by Vaccaro and others (2006) at 5-year intervals for 1985–2000 (shown in figure 21D). The 1985, 1990, and 1995 estimates from Vaccaro agree well with SOWAT results (RMSE = 0.028 MAF); however, the SOWAT estimate for 2000 was 0.11 MAF less than the estimate by Vaccaro and others (2006). These previous estimates also show a slight upward trend, which is not evident in the SOWAT irrigation

estimates. Irrigation applications modeled with SOWAT using actual ET to constrain climatic demand show larger inter-annual variability owing to climate and, in dry years, use of supplemental wells.

Groundwater pumping for irrigation generally is focused in areas where surface-water supplies are not available, although some of the pumping-center areas lie within or adjacent to surface-water irrigation districts. In 2007, these areas included parts of the Yakima, Pasco, Umatilla, and Walla Walla Basins and the Odessa area in western Adams County, Wash. (fig. 22). There also is a large component of smaller scale and more broadly distributed pumping for irrigation in areas like the Palouse Slope and adjacent areas in eastern Douglas, Lincoln, and Whitman Counties in Washington, and Wasco and Union Counties in Oregon (fig. 22).

Recharge in Irrigated Areas

Mean annual recharge in irrigated areas of the study area was 4.2 MAF (1985–2007) with 2.1 MAF (50 percent) occurring within the predominately surface-water irrigated regions of the Yakima Basin, Umatilla Basin, and CBP (fig. 23). Irrigation (5.3 MAF) is the largest source of water in these areas during the growing season, but precipitation contributes 4.4 MAF mostly during November–May. Annual recharge rates range from less than 5 in/yr in predominately sprinkler-irrigated areas where groundwater is the source to more than 20 in/yr in surface-water supplied areas where conveyance losses and less-efficient application methods are more common.

Model Limitations

SOWAT is a monthly water-balance model designed to take advantage of several readily available datasets that contain climate, soils, land-use, and water-use information in order to estimate the spatial and temporal variation in key components of the water budget for the CPRAS. SOWAT uses a fundamental representation of key processes in the soil zone with simplifying assumptions to estimate hydrologic flux and storage within the soil zone and between the soil zone and the saturated groundwater system. Model results, which include irrigation requirements and recharge, have associated uncertainties that are a product of both the simplifying assumptions and errors in the input data used by SOWAT. Examples of simplifying assumptions that may introduce model error include (1) that direct runoff is a constant fraction of precipitation and does not vary with precipitation amount, soil characteristics, or soil-moisture conditions, (2) that all soil moisture in excess of SMC becomes available for recharge and does not become runoff, and (3) that groundwater discharge contributes to ETa where precipitation and irrigation are insufficient. Each of these assumptions is justified based on its relative importance in the water budget and thus have limited contributions to overall model error.

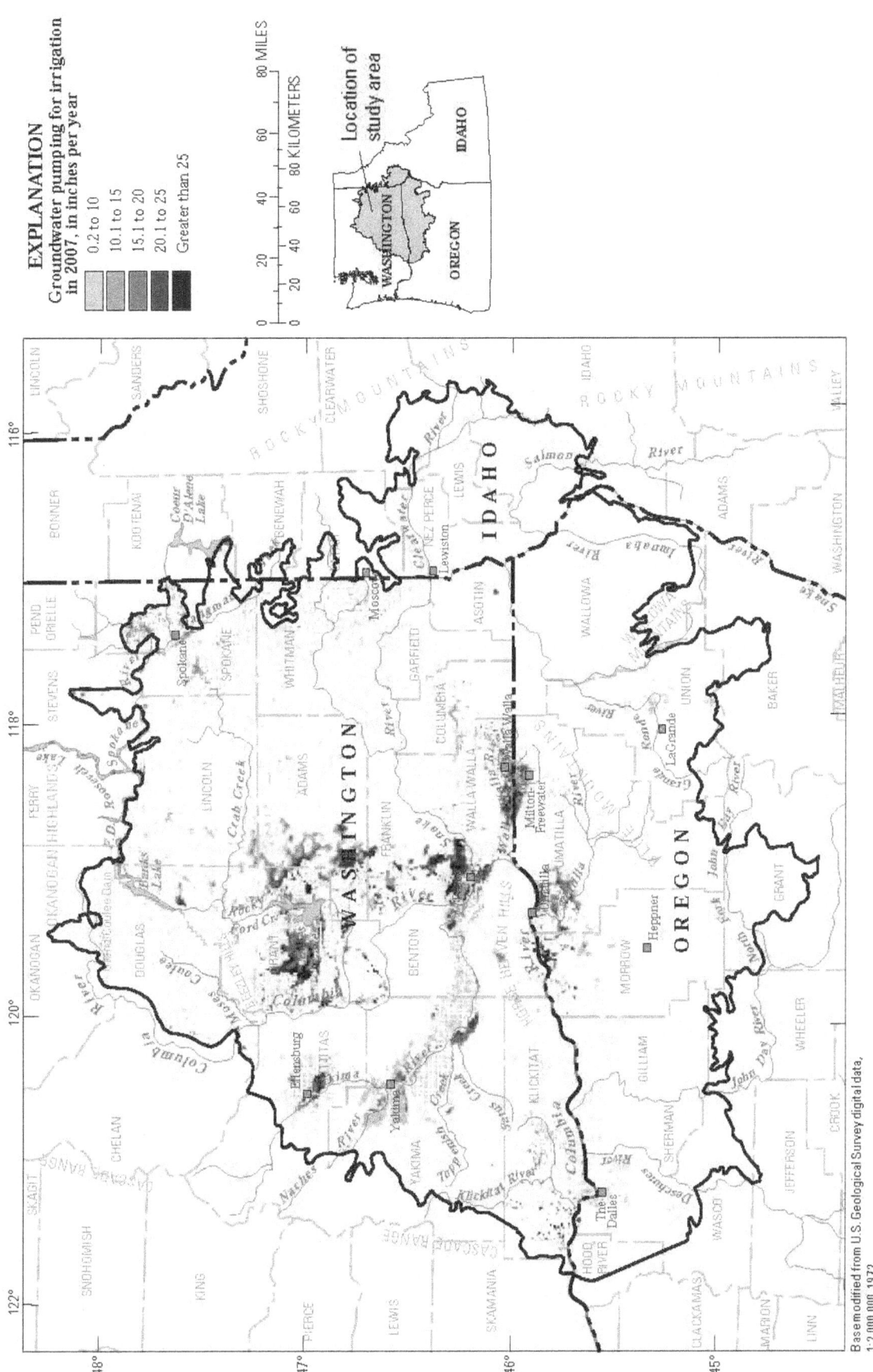

Figure 22. Distribution of groundwater withdrawals for irrigation in the Columbia Plateau Regional Aquifer System, 2007.

EXPLANATION

Groundwater pumping for irrigation
in 2007, in inches per year

0.2 to 10
10.1 to 15
15.1 to 20
20.1 to 25
Greater than 25

Location of
study area

Base modified from U.S. Geological Survey digital data,
1:2,000,000, 1972

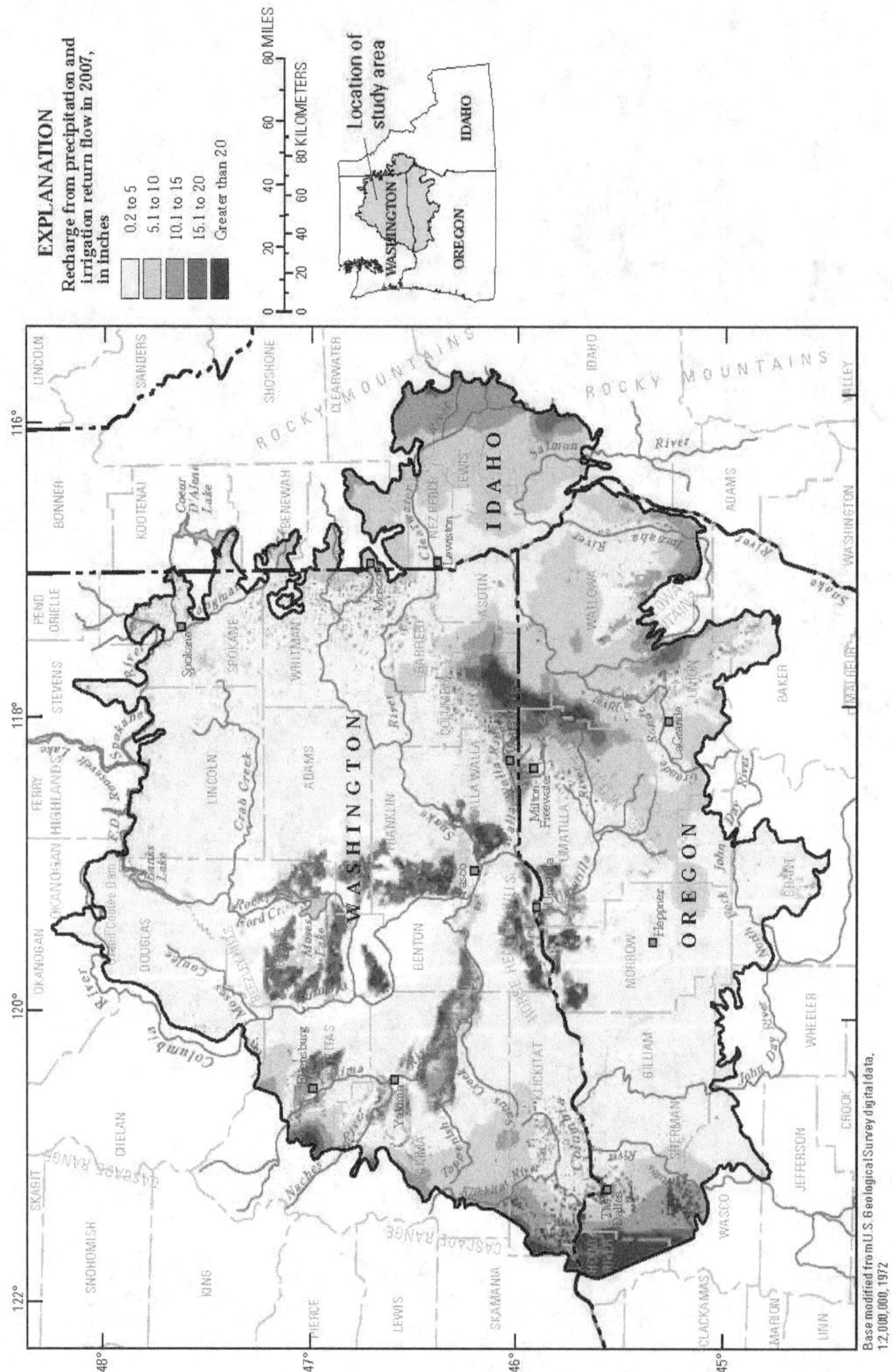

EXPLANATION

Recharge from precipitation and irrigation return flow in 2007, in inches

0.2 to 5
5.1 to 10
10.1 to 15
15.1 to 20
Greater than 20

Location of study area

Base modified from U.S. Geological Survey digital data,
1:2,000,000, 1972

Figure 23. Distribution of recharge from precipitation and irrigation return flow in the Columbia Plateau Regional Aquifer System, 2007.

Data-input errors are difficult to quantify, but some data types were developed using a wide variety of sources and assumptions and have inherently greater uncertainty. For example, IE is one of the most uncertain data inputs to the model and, because model results are highly sensitive to this parameter, is an area where significant improvements in estimates of irrigation could be made if better data were available. Specifically, if IE data could be obtained for and applied to subareas within the model, the likely variations that exist could be accounted for. Also, the fraction of irrigation supplied by groundwater is highly uncertain within and adjacent to some of the larger surface-water irrigation projects where groundwater is used as a supplemental source during dry periods. Precipitation and *ETa* are the hydroclimatic data that drive the model. These data, from PRISM and SSEB, respectively, are produced with documented methods and provide regional datasets with uncertainty that is appropriate for the scale of this analysis. If SOWAT were applied at smaller scales, the user might consider alternative sources of these data.

Non-Irrigation Groundwater Use

Non-irrigation groundwater uses in the CPRAS include public water supply, domestic, industrial, livestock, aquaculture, thermoelectric, and mining. Although small in comparison to the large amount of water used for irrigation (table 7), these uses are important to quantify. In order to estimate water use for an extensive study area with limited data, these water uses were aggregated into the following categories: public water supply, self-supplied domestic, industrial, and other uses. Annual totals of groundwater use for each of these categories were estimated for 1984–2009 and are shown in table 7. Groundwater-use estimates for the study area were made using a variety of sources, including information from USEPA's Safe Drinking Water Information System (SDWIS), U.S. Census Bureau population data (U.S. Census Bureau, 2010a, 2010b), published USGS State water-use estimates (Lane, 2009), and various State agencies (Curtis Stoehr, SDWIS Coordinator, Idaho Department of Environmental Quality, written commun., June 2010; Paul Cymbala, Drinking Water Program, Oregon Department of Human Services, written commun., June 2010).

Owing to the complexity in gathering comparable data across three States and the lack of measured pumpage values for the water-use categories, it was necessary to estimate groundwater use for the entire study area using a variety of different methods depending on the State, type of use, and availability of data.

Public Water Supply

For this study, PWS is defined as groundwater use withdrawn for human consumption by public and private water systems (cities, towns, rural water districts, and mobile-home parks) generally serving at least 25 people or having at least 15 connections. In Washington, public-supply systems are assigned to one of two categories: Group A or Group B. Group A systems are those that serve at least 25 people or have at least 15 connections, and Group B systems generally are smaller, serving less than 25 people per day or have less than 15 connections. For this study, Group B systems were not included in the public-supply estimate and are accounted for in the self-supplied domestic category described in the next section.

A list of Group A water systems (PWS) for the Washington State part of the study area was obtained from the Washington State Department of Health–Office of Drinking Water (Washington State Department of Health, 2010). The dataset included information such as type of system (community or non-community), system characteristics (residential, commercial, school etc.), population served, number of connections, source type (well, spring, or surface water), effective and inactive dates for the PWS system, and for the facilities (wells).

The USGS publishes water-use estimates, by category, for each State at 5-year intervals (U.S. Geological Survey, 2010). These compilations contain county-level domestic and public-supply per capita water-use rates in gallons per day. Public supply and domestic per capita water-use rates used in this study, 282 and 168 gal/d, respectively, were calculated by taking the average rate for each of these two categories for all study area counties from the 1985, 1990, 1995, 2000, and 2005 compilations. For comparison, the reported range of PWS rates in 2000 for all counties within the study area is 148 gal/d in Jefferson County, Ore. and 765 gal/d in Grant County, Oreg.

For PWS community systems with a principal-system characteristic of "residential," water-use estimates were made by multiplying the population served by the average domestic per capita rate of 168 gal/d. Any community wells with a principal characteristic of residential that were outside of incorporated areas were removed from the PWS dataset to avoid duplication with the domestic self-supplied water-use estimate. For all other community systems, the population served was multiplied by the average PWS per capita rate of 282 gal/d. For non-community systems without published population-served numbers, the total number of connections was multiplied by an average-persons per connection of 2.45 and then by the PWS per capita rate. The average-persons per connection was calculated by dividing the population served by the total connections of the community systems and then averaging the result.

Table 7. Estimated annual groundwater use for the Columbia Plateau Regional Aquifer System, 1984–2009.

[Estimates of irrigation groundwater use were not made for 1984, 2008, and 2009; groundwater use in acre-feet. **Abbreviation:** NA, not applicable]

	1984	1985	1986	1987	1988	1989	1990	1991	1992
Irrigation	NA	1,391,000	1,382,000	1,615,000	1,514,000	1,278,000	1,392,000	1,685,000	1,312,000
Public supply	200,600	201,500	200,400	200,400	200,700	201,800	205,100	209,400	215,700
Self-supplied domestic	54,580	54,650	54,280	54,200	54,340	54,750	55,530	56,870	58,340
Industrial	53,390	53,390	52,430	51,470	50,510	49,550	48,580	50,870	53,160
Other	16,900	16,900	17,140	17,380	17,610	17,850	19,860	18,830	18,320
Total annual pumpage	NA	1,717,000	1,706,000	1,938,000	1,837,000	1,602,000	1,721,000	2,021,000	1,658,000

	1993	1994	1995	1996	1997	1998	1999	2000	2001
Irrigation	1,179,000	1,650,000	1,425,000	1,315,000	1,413,000	1,072,000	1,092,000	1,503,000	1,221,000
Public supply	221,200	225,800	230,700	236,900	239,100	241,500	244,300	245,700	248,000
Self-supplied domestic	59,920	61,350	62,560	63,330	64,040	64,740	65,360	65,820	66,160
Industrial	55,440	57,730	60,020	58,650	57,280	55,910	54,550	53,180	51,330
Other	17,810	17,310	20,440	20,550	20,660	20,760	20,870	22,950	25,360
Total annual pumpage	1,533,000	2,012,000	1,799,000	1,694,000	1,794,000	1,455,000	1,477,000	1,891,000	1,612,000

	2002	2003	2004	2005	2006	2007	2008	2009	Average annual pumpage
Irrigation	1,322,000	1,436,000	1,175,000	1,417,000	1,554,000	2,088,000	NA	NA	1,410,000
Public supply	251,500	253,700	256,200	256,400	259,500	262,800	267,300	269,100	232,500
Self-supplied domestic	66,750	67,350	68,000	68,580	69,320	70,090	71,160	71,160	62,430
Industrial	49,480	47,630	45,780	43,930	43,930	43,930	43,930	43,930	51,150
Other	27,780	30,200	32,620	43,610	43,610	43,610	43,610	43,610	25,240
Total annual pumpage	1,718,000	1,835,000	1,578,000	1,830,000	1,970,000	2,508,000	NA	NA	1,781,000

The reported source-effective date of the well was used to represent the start date of pumpage, and the reported source-inactive date was used to represent the end date of pumpage. For permanent wells, the daily rate was multiplied by 365.2 to obtain yearly estimates, and for seasonal wells it was assumed that the well was pumped for 90 days. Population-served values for the systems were assumed to be current for 2009 and were adjusted using the U.S. Census Bureau intercensal population-change estimates by county (U.S. Census Bureau, 2010b).

PWS water-use estimates for the study area counties in Idaho and Oregon were made using information queried from USEPA's SDWIS by the Idaho Department of Environmental Quality and the Oregon Office of Environmental Public Health. SDWIS contains information about public water systems including type of system (community or non-community), population served, number of connections, source type (well, spring, or surface water), and effective and inactive dates for the facilities (wells) as well as violations of USEPA's drinking-water regulations, as reported to USEPA by individual States. The methods used to calculate public supply water use for Idaho and Oregon were identical to those used for Washington, with one exception. Because there was no SDWIS information about system characteristics (residential, commercial, school etc.), estimates for all the systems were made by multiplying the PWS per capita rate times the population served or total connections; the average domestic rate was not used to adjust for residential use.

Self-Supplied Domestic

For this study, self-supplied domestic water use refers to groundwater withdrawn for indoor and outdoor single-household use outside of incorporated areas. Owing to the lack of information about exempt wells (wells not required to go through the permitting process, primarily used for self-supplied domestic use, and use less than 5,000, 15,000, and 13,000 gal/d, for Washington, Oregon, and Idaho, respectively), estimates for this category were made using the 168 gal/d per capita rates and census-block populations for the unincorporated parts of the study area.

To estimate the domestic self-supplied population, U.S. Census Bureau Census 2000 County Demographics (SF1) tables for Washington, Oregon, and Idaho were joined with Census 2000 TIGER®/Line shapefiles for each State (U.S. Census Bureau, 2010a). Census blocks are geographic subdivisions of census-block groups and are the smallest geographic area for which the U.S. Census Bureau assembles

and tabulates data. The TIGER®/Line shapefiles are a digital database of geographic features, including census-block polygons that have unique identifiers (STFIDs) for each block that can be used to join the demographics tables to the shapefiles. Most of the population in incorporated areas are served by PWS systems, therefore, census blocks in incorporated areas were removed. In order to ensure a complete accounting, all census blocks that were within the study area and those that were within 1 mi of the study area boundary were included.

To account for population changes during the study period, U.S. Census Bureau county-level intercensal estimates were used to determine an annual percent change in population that could be applied to each census block (U.S. Census Bureau, 2010b). For comparison, the range in population change from 2000 to 2001 was -3.9 percent for Garfield County, Wash., and 2.9 percent for Franklin County, Wash. To compute a final domestic self-supplied water use, the average per capita rate of 168 gal/d was multiplied by the estimated population of each census block for every year of the study period. The distribution and range of values are shown in figure 24.

Industrial

Industrial water use includes self-supplied groundwater withdrawn for fabricating, processing, and manufacturing of a product. Total industrial water use is reported in million gallons per day for each county as part of the USGS State water-use estimates (U.S. Geological Survey, 2010). To estimate industrial groundwater use for the study area counties, total industrial groundwater amounts for each 5-year interval were extracted from the published USGS estimates, and a linear interpolation was used to approximate the values for years in between compilations.

Other Uses

Owing to data gaps for the study area and time period, an estimate of all other groundwater uses was made by combining available USGS State groundwater-use totals (in million gallons per day (Mgal/d)) in the study area counties for the following categories: mining, thermoelectric, livestock, and aquaculture. As with industrial groundwater use, a linear interpolation was used to calculate an approximate amount of groundwater use for years in between the 5-year compilations. Estimates were not made for thermoelectric and aquaculture until the 1995 compilation.

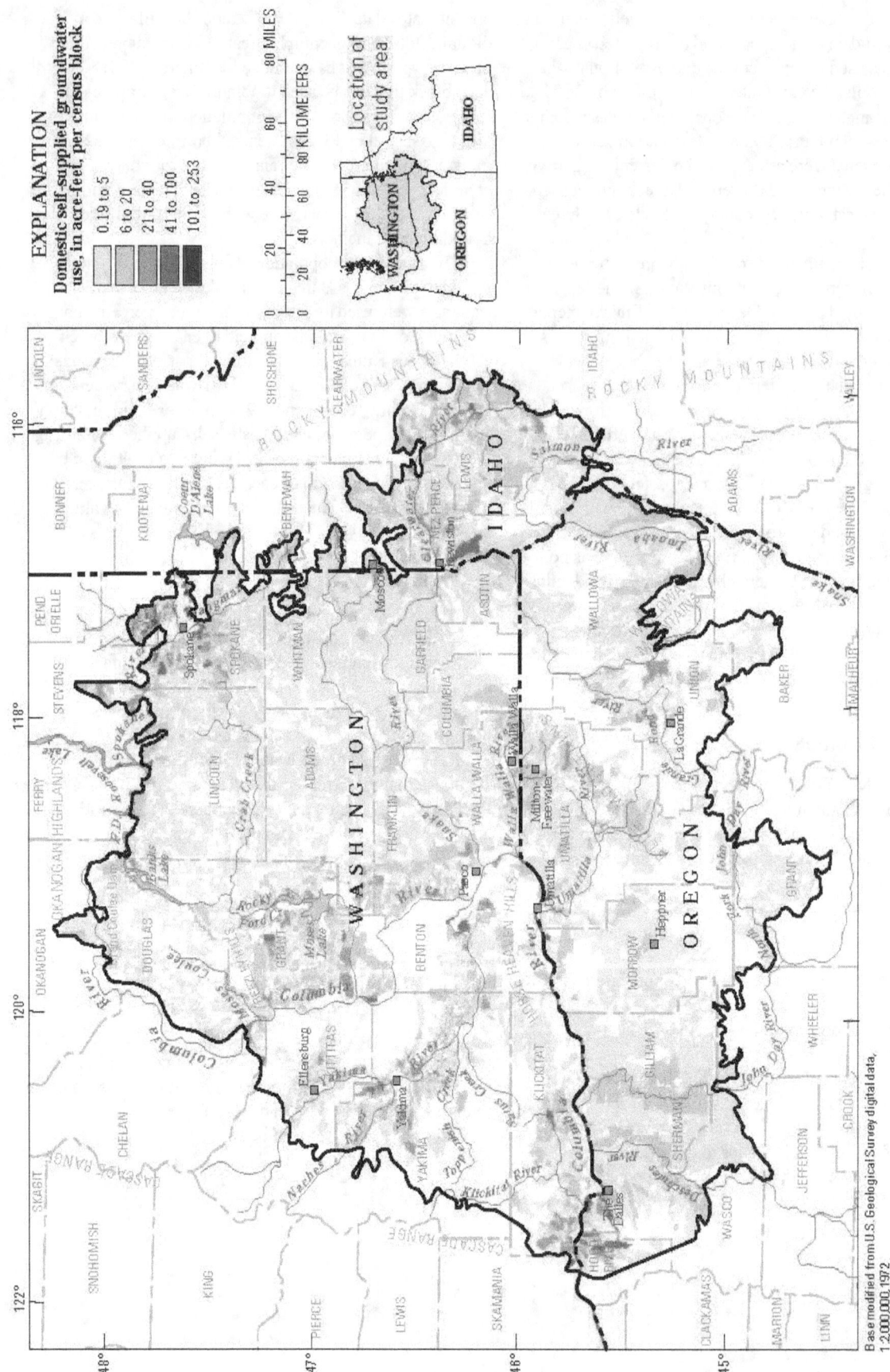

Base modified from U.S. Geological Survey digital data,
1:2,000,000, 1972

Figure 24.   Distribution of estimated domestic self-supplied groundwater use by census blocks in the Columbia Plateau Regional Aquifer System, 2000.

Results

The groundwater use estimated in the study area for the above categories for 1984–2009 is shown in table 7. The estimated annual groundwater use shown in figure 25 clearly shows a large change over time.

Public Water Supply

PWS groundwater use has increased approximately 34 percent from 200,600 acre-ft/yr in 1984 to 269,100 acre-ft/yr in 2009 (table 7). This increase is due to the increased population (fig. 26) used in the water-use calculations and by the addition of new public-supply wells during the study period. Owing to source-water security concerns, locations of public-supply wells are not presented spatially. In the Columbia Plateau, the four counties with the largest amount of PWS groundwater use in ascending order are Grant, Benton, Yakima, and Spokane (all in Washington State).

Domestic

There was an approximate 30-percent increase in domestic self-supplied groundwater use in the CPRAS from 54,580 acre-ft/yr in 1984 to 71,160 acre-ft/yr in 2009 (table 7, fig. 25). Because the domestic groundwater-use estimation is based on census-block population, the 2000 spatial distribution of water use reflects spacial variations in population (fig. 24).

Industrial

According to USGS estimates (U.S. Geological Survey, 2010), industrial groundwater use in the Columbia Plateau has decreased from approximately 53,390 acre-ft/yr in 1984 to 43,930 acre-ft/yr in 2009. There are several explanations for this decline including a decrease in industrial-production hours during the past 15 years, the adoption of more water-efficient processes, and a shift in the type of industry from those that require large amounts of water (refining and manufacturing of wood products) to ones that use less water (computer and electronic manufacturing) (R.C. Lane, U.S. Geological Survey, written commun., 2010).

Other

According to USGS estimates (U.S. Geological Survey, 2010), groundwater use for mining, thermoelectric, livestock, and aquaculture in the Columbia Plateau has increased from approximately 16,900 acre-ft/yr in 1984 to 43,610 acre-ft/yr in 2009. The main reason for this increase is owing to the increase in population and the resulting demand for products from each category (R.C. Lane, U.S. Geological Survey, written commun., 2011). However, during the study period (1984–2009), the USGS 5-year water-use compilations have varied in data requirements, collection methods, and data sources; therefore, it is difficult to assess and compare groundwater use for these categories.

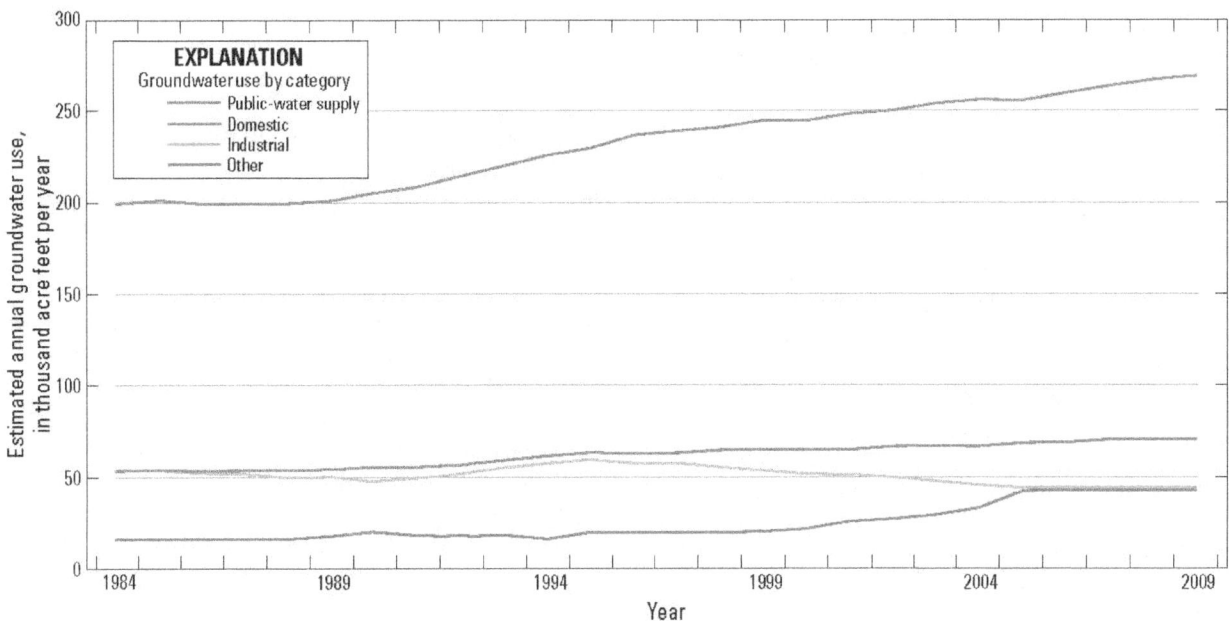

Figure 25. Estimated annual groundwater use for public water supply, domestic, industrial, and other uses, Columbia Plateau Regional Aquifer System, 1984–2009.

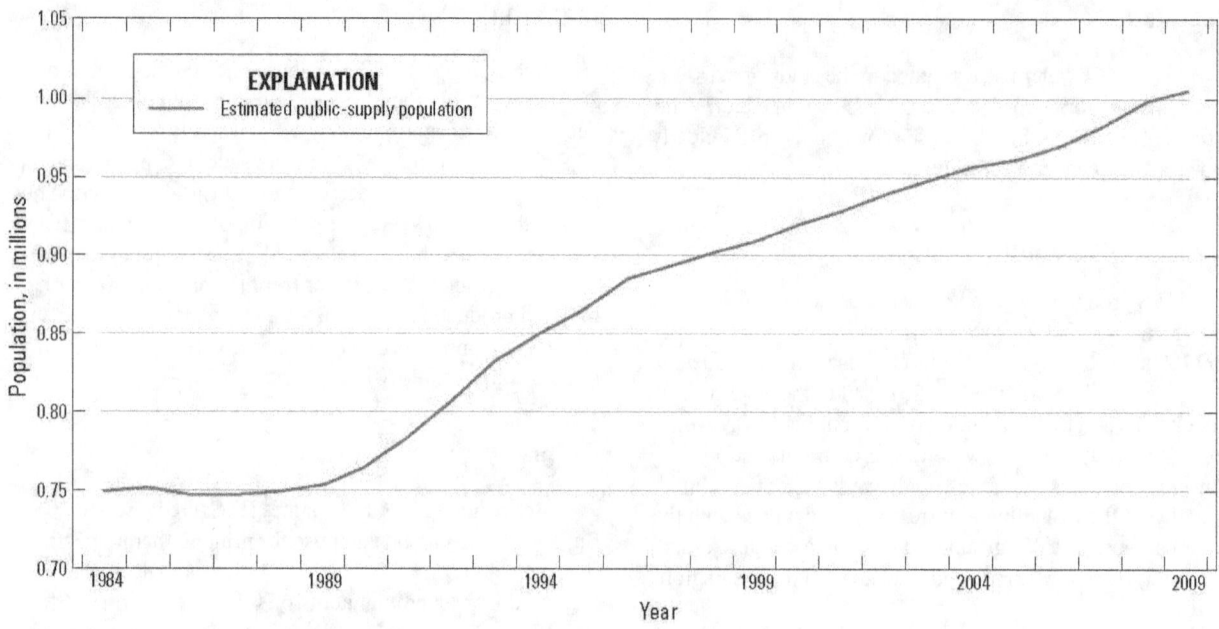

Figure 26.   Estimated public water-supply population for the Columbia Plateau Regional Aquifer System, 1984–2009.

Summary and Conclusions

The Columbia Plateau is located in the northwest part of the United States. It is bound by the Rocky Mountains to the east, the Cascade Range to the west, and the Okanogan Highlands on the north. The southern boundary is defined as the mapped extent of the Columbia River Basalt Group (CRBG). The Columbia Plateau Regional Aquifer System (CPRAS) covers an area of about 44,000 square miles in a structural and topographic basin within the drainage of the Columbia River. The primary aquifers within the Columbia Plateau occur in basalts of the CRBG and in places, overlying sediment. Conceptually, the system is a series of productive basalt aquifers consisting of permeable interflow zones separated by less permeable flow interiors, overlaid locally by aquifers composed of sedimentary material.

Much of the area is semi-arid, with precipitation as low as 6 inches per year (in/yr) in the lower-altitude parts of the basin where natural vegetation is principally sagebrush and grasslands. Precipitation increases to more than 45 in/yr in the higher altitude forested areas. Of 2 million acres irrigated, one-third is supplied by groundwater. Approximately 80 percent of the entire groundwater use in the study area is for irrigation in support of the $6 billion per year agricultural economy. Groundwater issues in the Columbia Plateau include competing agricultural, domestic, and environmental demands.

The CPRAS includes seven hydrogeologic units—the Overburden unit, three units in the permeable basalt rock, two confining units, and the basement confining unit. The three basalt units are the Saddle Mountains, Wanapum, and Grande Ronde Basalts and their intercalated sediments. Median thickness of the Saddle Mountains and Wanapum units are 280 and 330 feet, respectively. Median thickness of the deepest and most-voluminous unit, the Grande Ronde unit, is largely unknown, but total thickness may exceed 14,000 feet near the center of the basin. The confining units are equivalent to the Saddle Mountains-Wanapum and Wanapum-Grande Ronde interbeds, referred to in this study as the Mabton and Vantage Interbeds, respectively. The interbed units are fairly extensive laterally, but are thin (generally tens of feet), especially when compared with the thickness of the basalt units. The basement confining unit, referred to as Older Bedrock, consists of pre-CRBG rocks that generally have much lower permeabilities than the basalts and are considered the base of the regional flow system.

Lateral hydraulic conductivity (K_h) of the hydrogeologic units ranges widely, indicating the heterogeneity of the geologic materials making up the aquifer system. Average or effective K_h values of the water-producing zones in the Overburden unit are on the order of 1 to 800 feet per day (ft/d) and are about 1 to 5 ft/d for the CRBG units. Effective K_h for the Older Bedrock unit appears to be about 0.0001 to 3 ft/d.

Based on specific-capacity data, mean and median values for the Overburden unit was 1,694 and 161 ft/d, respectively. The specific-capacity derived mean and median values for the basalt units were 805 and 70 ft/d, respectively, and for the Older Bedrock unit they were 46 and 6 ft/d, respectively. Vertical hydraulic conductivity (K_v) of the units largely is unknown. K_v values have been estimated to range from about 0.009 to 2 ft/d for the Overburden unit; values for the clay-to-shale parts of this unit may be as small as 10^{-10} to 10^{-7} ft/d. Reported K_v values for the CRBG units range from 4×10^{-7} to 4 ft/d.

Within the basalt units, groundwater moves both laterally and vertically in the basalt interflow zones, flow centers, and sedimentary interbeds. Lateral hydraulic conductivities generally are greatest in the interflow zones. Hydraulic conductivities in the flow centers are controlled by secondary features—the predominantly vertical joints and fractures of the entablature and colonnade. Consequently, the interflow zones support most of the lateral groundwater movement, whereas movement in the flow center mainly is vertical. Therefore, the interflow zones in the basalt sequence are numerous, thin, semiconfined aquifers whose physical and hydraulic characteristics vary laterally and vertically.

Except for groundwater flow in the deeply buried parts of the system, large-scale structural control compartmentalizes the flow system in places. The compartmentalization limits the length of the flow paths, resulting in relatively short paths for such a large aquifer system. Structural control is exerted primarily by the major ridges in the Yakima Fold Belt. Groundwater levels in the basalt units generally mimic the shape of land surface or, where a unit is buried, the dip of the basalt, because most groundwater occurs and moves in the interflow zones. Where the units are deeply buried, water-level contours are smoother owing to lower hydraulic gradients than those in the uplands, which typically are outcrop areas.

Selected carbon-14 and oxygen-18 (^{14}C and δ^{18}O) data were compiled to demonstrate some of the general patterns of groundwater age that are found in the CPRAS. A small group of post-bomb samples—those with ^{14}C concentrations greater than 100 percent modern carbon (PMC) represent modern recharge and are from shallow wells. Two groups of low-^{14}C samples are evident: those with ^{14}C concentrations of less than 20 PMC and δ^{18}O values less than (more negative than) -17.6 units of per mil (‰) (ranging from -19.2 to -17.7‰), and those with ^{14}C concentrations of less than 20 PMC but δ^{18}O values greater than -17.5‰ (ranging from -17.4 to -15.8‰). The group of low-^{14}C samples that have low δ^{18}O values likely represent Pleistocene groundwater because not only are the ^{14}C concentrations less than 20 PMC, but the magnitude of the depletion in ^{18}O is consistent with recharge during the Pleistocene. As a group, these tend to be the deepest of the wells with data compiled for this analysis. Other samples could represent groundwater with ages intermediate between Pleistocene and the present (post-1950), and some could represent mixtures of Pleistocene and late Holocene or modern groundwater.

Recharge from infiltration of precipitation was estimated for the CPRAS using a polynomial-regression equation based on annual precipitation and the results of recharge modeling done in the 1980s. Annual-recharge distributions were computed for 23 years (1985–2007) using the regression equation with gridded annual-precipitation data. In the CPRAS, the mean annual recharge from infiltration of precipitation was estimated to be 4.6 in/yr (14,980 cubic feet per second). The spatial distribution in recharge also mirrors that of annual precipitation, with the highest recharge (more than 20 in/yr) occurring in the upper Yakima Basin, the Blue Mountains southeast of Walla Walla, and adjacent to the Columbia River where it leaves the study area through the gorge in the Cascade Range. Mean annual recharge from infiltration of precipitation is less than 1 in/yr for a large portion of the study area adjacent to the Columbia and Yakima Rivers where precipitation is limited to less than 10 in/yr.

A regional-scale hydrologic budget was developed using a monthly SOil WATer balance (SOWAT) model to estimate irrigation water demand, historically one of the largest and most difficult parts of the water budget to quantify. The SOWAT model also was used to estimate groundwater flux (recharge or discharge), direct runoff, and soil moisture within irrigated areas. The SOWAT model was specifically developed to make use of estimates of actual evapotranspiration (ET) available from a new Simplified Surface Energy Balance method that utilizes remotely sensed land-surface temperature data. The resulting soil-water balance in irrigated lands is dominated by the climatic water demand of ET and the application of irrigation water to satisfy crop-water requirements. Precipitation replenishes soil moisture and contributes to groundwater recharge in the late fall and winter months (November–January). Some early spring ET demand is supplied by soil moisture, but beginning in April, irrigators apply sufficient water to maintain soil moisture above the maximum allowable depletion. Groundwater recharge in the irrigated lands peaks again in the summer with ET demand and irrigation applications and tapers in the fall before precipitation increases. Mean monthly irrigation throughout the study area peaks in July at 1.6 million acre-ft (MAF) (1985–2007 average), of which 0.45 and 1.15 MAF are from groundwater and surface-water sources, respectively. Annual irrigation water use in the study area averaged 5.3 MAF during 1985–2007, with 1.4 MAF (or 26 percent) supplied from groundwater and 3.9 MAF supplied from surface water. Mean annual recharge from irrigation return flow in the study area was 4.2 MAF (1985–2007) with 2.1 MAF (50 percent) occurring within the predominately surface-water irrigated regions of the Yakima Basin, Umatilla Basin, and Columbia Basin Project. Annual recharge rates range from less than 5 in/yr in predominately sprinkler-irrigated areas where groundwater is the source to more than 20 in/yr in surface-water supplied areas where conveyance losses and less-efficient application methods are more common.

In addition to irrigation, annual groundwater use (1984–2009) was estimated for the following categories: public supply, self-supplied domestic, industrial, and other uses. Groundwater-use estimates for the study area were made using a variety of sources, including information from the U.S. Environmental Protection Agency's Safe Drinking Water Information System, U.S. Census Bureau population data, published USGS State water-use estimates, and various State agencies. Public-supply groundwater use within the study area increased from 200,600 acre-ft/yr in 1984 to 269,100 acre-ft/yr in 2009. Domestic self-supplied groundwater use increased from 54,580 acre-ft/yr in 1984 to 71,160 acre-ft/yr in 2009. Industrial groundwater use decreased from 53,390 acre-ft/yr in 1984 to 43,930 acre-ft/year in 2009 and likely is attributable to a decrease in industrial-production hours during the past 15 years; the adoption of more water-efficient processes; and a shift in the type of industry from those that require large amounts of water (refining and manufacturing of wood products) to ones that use less water (computer and electronic manufacturing). Other groundwater use, including that used for mining, thermoelectric needs, livestock, and aquaculture combined, increased from 16,900 acre-ft/yr in 1984 to 43,610 acre-ft/yr in 2009.

Acknowledgments

The authors acknowledge Curtis Stoehr, SDWIS Coordinator of the Idaho Department of Environmental Quality–State Office, for conducting a query of the SDWIS for Idaho. The authors also acknowledge Joe Carlson (Manager) and Paul Cymbala, Office of Environmental Public Health, State Drinking Water Program for Oregon, for conducting a query of the SDWIS for Oregon.

References Cited

Allen, R.G., Tasumi, M., and Trezza, R., 2007, Satellite-based energy balance for mapping evapotranspiration with internalized calibration (METRIC)—Applications: American Society of Civil Engineers, Journal of Irrigation and Drainage Engineering, v. 133, no. 4, p. 380–394.

Alley, W.M., 1984, On the treatment of evapotranspiration, soil moisture accounting, and aquifer recharge in monthly water balance models: Water Resources Research, v. 20, p. 1137–1149.

Alley, W.M., 1985, Water balance models in one-month-ahead streamflow forecasting: Water Resources Research, v. 21, p. 597–606.

Ames, L.L., 1980, Hanford basalt flow mineralogy: Richland, Wash., Pacific Northwest Laboratory, PNL-2847, 447 p.

Bauder, J., and Carlson, L., 2010, Soil moisture depletion—What does 65% depletion mean?: Montana State University Water Quality and Irrigation Management, accessed June 15, 2011, at http://waterquality.montana.edu/docs/irrigation/depletion.shtml.

Bauer, H.H., and Hansen, A.J., 2000, Hydrology of the Columbia Plateau aquifer system, Washington, Oregon, and Idaho: U.S. Geological Survey, Water-Resources Investigations Report 96–4106, 61 p. (Also available at http://pubs.er.usgs.gov/usgspubs/wri/wri964106.)

Bauer, H.H., and Vaccaro, J.J., 1987, Documentation of a Deep Percolation Model for estimating ground-water recharge: U.S. Geological Survey Open-File Report 86–536, 180 p. (Also available at http://pubs.er.usgs.gov/usgspubs/ofr/ofr86536.)

Bauer, H.H., and Vaccaro, J.J., 1990, Estimates of ground-water recharge to the Columbia Plateau regional aquifer system, Washington, Oregon, and Idaho, for predevelopment and current land use conditions: U.S. Geological Survey Water-Resources Investigations Report 88–4108, 37 p. (Also available at http://pubs.er.usgs.gov/publication/wri884108.)

Bauer, H.H., Vaccaro, J.J., and Lane, R.C., 1985, Maps showing ground-water levels in the Columbia River Basalt Group and overlying material, spring 1983, southeastern Washington: U.S. Geological Survey, Water-Resources Investigations Report 84–4360, 4 pls. (Also available at http://pubs.er.usgs.gov/usgspubs/wri/wri844360.)

Benson, L.V., and Teague, L.S., 1982, Diagenesis of basalts from the Pasco basin, Washington—Distribution and composition of secondary mineral phases: Journal of Sedimentary Petrology, v. 52, no. 2, p. 595–613.

Bjornstad, B.N., Babcock, R.S., and Last, G.V., 2007, Flood basalts and Ice Age floods—Repeated late Cenozoic cataclysms of southeastern Washington, in Stelling, P., and Tucker, D.S., eds., Floods, Faults, and Fire—Geological Field Trips in Washington State and Southwest British Columbia: Geological Society of America Field Guide 9, p. 209–255.

Blaney, H.F., and Criddle, W.D., 1950, Determining water requirements in irrigated areas from climatological and irrigation data: U.S. Soil Conservation Service, Technical Paper 96, 48 p.

Blaser, P.C., Coetsiers, M., Aeschbach-Hertig, W., Kipfer, R., Van Camp, M., Loosli, H.H., and Walraevens, K., 2010, A new groundwater radiocarbon correction approach accounting for palaeoclimate conditions during recharge and hydrochemical evolution—The Ledo-Paniselian Aquifer, Belgium: Applied Geochemistry, v. 25, p. 437–455.

Bolke, E.L., and Skrivan, J.A., 1981, Digital-model simulation of the Toppenish alluvial aquifer, Yakima Indian Reservation, Washington: U.S. Geological Survey Open-File Report 81–425, 34 p.

Bredehoeft, J.D., Neuzil, C.E., and Milly, P.C.D, 1983, Regional flow in the Dakota aquifer—A study of the role of confining layers: U.S. Geological Survey Water Supply Paper 2237, 34 p. (Also available at http://pubs.er.usgs.gov/usgspubs/wsp/wsp2237.)

Brown, K.B., McIntosh, J.C., Baker, V.R., and Gosch, Damian, 2010, Isotopically-depleted late Pleistocene groundwater in Columbia River Basalt aquifers—Evidence for recharge of glacial Lake Missoula floodwaters?: Geophysical Research Letters, v. 37, L21402, 5 p., doi:10.1029/2010GL044992, accessed June 15, 2011, at http://www.agu.org/journals/gl/gl1021/2010GL044992/.

Burns, E.R., Morgan, D.S., Peavler, R.S., and Kahle, S.C., 2011, Three-dimensional model of the geologic framework for the Columbia Plateau Regional Aquifer System, western Idaho, northeastern Oregon, and southeastern Washington: U.S. Geological Survey Scientific Investigations Report 2010–5246, 44 p. (Also available at http://pubs.usgs.gov/sir/2010/5246/.)

Bush, John, Gill, Steve, Petrich, Christian, and Pierce, Jack, 1999, Geological and hydrogeological references—Palouse Region: University of Idaho, Palouse Basin Aquifer Committee Technical Report 99-02, accessed June 15, 2011, at http://www.webs.uidaho.edu/pbac/pubs/biblio99.pdf.

Carnahan, B., Luther, H.A., and Wilkes, J.O., 1969, Applied numerical methods: New York, Wiley, 604 p.

Clark, I.D., and Fritz, Peter, 1997, Environmental isotopes in hydrogeology: New York, Lewis Publishers, 328 p.

Cleveland, W.S., Grosse, E., and Shyu, W.M., 1992, Local regression models, chap. 8, in Chambers, J.M., and Hastie, T.J., eds., Statistical models in S: Pacific Grove, Calif., Wadsworth & Brooks/Cole Advanced Books & Software, p. 309–376.

Conlon, T., 2006, Columbia River basalt stratigraphy in the Pacific Northwest: U.S. Geological Survey, accessed June 16, 2011, at http://or.water.usgs.gov/projs_dir/crbg/.

Conlon T.D., Wozniak, K.C., Woodcock, D., Herrera, N.B., Fisher, B.J., Morgan, D.S., Lee, K.K., and Hinkle, S.R., 2005, Ground-water hydrology of the Willamette Basin, Oregon: U.S. Geological Survey Scientific Investigations Report 2005–5168, 83 p.

Converse Consultants NW, 1991, Construction and testing of Kissell Park well for City of Yakima, Washington: Converse Consultants, Draft report prepared for R.W. Beck and Associates, October 1989.

Coplen, T.B., 1994, Reporting of stable hydrogen, carbon, and oxygen isotopic abundances: Pure and Applied Chemistry, v. 66, p. 273–276.

Davies-Smith, A., Bolke, E.L., and Collins, C.A., 1988, Geohydrology and digital simulations of the ground-water flow system in the Umatilla Plateau and Horse Heaven Hills area, Oregon, and Washington: U.S. Geological Survey, Water-Resources Investigations Report 87–4268, 77 p. (Also available at http://pubs.er.usgs.gov/djvu/WRI/wrir_87_4268.djvu.)

Douglas, A.A., Osiensky, J.L., and Keller, C.K., 2007, Carbon-14 dating of ground water in the Palouse Basin of the Columbia River basalts: Journal of Hydrology, v. 334, p. 502–512.

Drost, B.W., Cox, S.E., and Schurr, K.M., 1997, Changes in ground-water levels and ground-water budgets, from predevelopment to 1986, in parts of the Pasco basin, Washington: U.S. Geological Survey Water-Resources Investigations Report 96–4086, 172 p. (Also available at http://pubs.er.usgs.gov/usgspubs/wri/wri964086.)

Drost, B.W., Whiteman, K.J., and Gonthier, J.B., 1990, Geologic framework of the Columbia Plateau Aquifer System, Washington, Oregon, and Idaho: U.S. Geological Survey Water-Resources Investigations Report 87–4238, 13 p. (Also available at http://pubs.er.usgs.gov/djvu/WRI/wrir_87_4238.djvu.)

Ellis, A.W., Hawkins, T.W., Balling, R.C., and Gober, P., 2008, Estimating future runoff levels for a semi-arid fluvial system in central Arizona, USA: Climate Research, v. 35, no. 3, p. 227–239.

Ferris, J.G., Knowles, D.B., Brown, R.H., and Stallman, R.W., 1962, Theory of aquifer tests: U.S. Geological Survey Water Supply Paper 1536-E, 174 p. (Also available at http://pubs.er.usgs.gov/usgspubs/wsp/wsp1536E.)

Freeze, R.A., and Cherry, J.A., 1979, Groundwater: Englewood Cliffs, N.J., Prentice-Hall, 604 p.

Gannett, M.W., and Lite, K.E., 2004, Simulation of regional ground-water flow in the upper Deschutes basin, Oregon: U.S. Geological Survey Water-Resources Investigations Report 03–4195, 84 p. (Also available at http://pubs.er.usgs.gov/usgspubs/wri/wri034195.)

Gannett, M.W., Lite, K.E., Morgan, D.S., and Collins, C.A., 2001, Ground-water hydrology of the Upper Deschutes Basin, Oregon: U.S. Geological Survey Water-Resources Investigations Report 2000–4162, 77 p.

Golder Associates, 2002, Naches Basin (WRIA 38) storage assessment application of aquifer storage and recovery: Golder Associates, File no. 983-1085-002.400, 48 p. (plus figures and tables)

Golder Associates, 2004, Phase II—Level 1 Technical Assessment for the Palouse Basin (WRIA 34) Watershed Planning Unit: Golder Associates, File no. 043-1064.1140 [variously paged].

Gowda, P.H., Senay, G.B., Howell, T.A., and Marek, T.H., 2009, Lysimetric evaluation of the Simplified Surface Energy balance approach in the Texas High Plains: Applied Engineering in Agriculture, v. 22, no. 5, p. 665–669.

Hamon, W.R., 1961, Estimating potential evapotranspiration: Journal of the Hydraulics Division, Proceedings of the American Society of Civil Engineers, v. 87, p. 107–120.

Hansen, A.J., Vaccaro, J.J., and Bauer, H.H., 1994, Ground-water flow simulation of the Columbia Plateau Regional Aquifer System, Washington, Oregon, and Idaho: U.S. Geological Survey Water-Resources Investigations Report 91–4187, 101 p., 15 pls. (Also available at http://pubs.er.usgs.gov/usgspubs/wri/wri914187.)

Hearn, P.P., Steinkampf, W.C., Bortleson, G.C., and Drost, B.W., 1985, Geochemical controls on dissolved sodium in basalt aquifers of the Columbia Plateau, Washington: U.S. Geological Survey Water-Resources Investigations Report 84–4304, 38 p., 1 pl. (Also available at http://pubs.er.usgs.gov/usgspubs/wri/wri844304.)

Hinkle, S.R., 1996, Age of ground water in basalt aquifers near Spring Creek National Fish Hatchery, Skamania County, Washington: U.S. Geological Survey Water-Resources Investigations Report 95–4272, 26 p. (Also available at http://pubs.er.usgs.gov/djvu/WRI/wrir_95_4272.djvu.)

Hotchkiss, W.R., and Levings, J.F., 1986, Hydrogeology and simulation of water flow in strata above Bearpaw Shale and equivalents of eastern Montana and northeastern Wyoming: U.S. Geological Survey Water-Resources Investigations Report 85–4281, 72 p. (Also available at http://pubs.er.usgs.gov/usgspubs/wri/wri854281.)

Jones, M.A., and Vaccaro, J.J., 2008, Extent and depth to top of basalt and interbed hydrogeologic units, Yakima River Basin Aquifer System, Washington: U.S. Geological Survey Scientific Investigations Report 2008–5045, 32 p., 5 pls. (Also available http://pubs.usgs.gov/sir/2008/5045/pdf/sir20085045.pdf.)

Jones, M.A., Vaccaro, J.J., and Watkins, A.M., 2006, Hydrogeologic framework of sedimentary deposits in six structural basins, Yakima River Basin, Washington: U.S. Geological Survey Scientific Investigations Report 2006–5116, 24 p. (Also available at http://pubs.usgs.gov/sir/2006/5116/section4.html.)

Kahle, S.C., Olsen, T.D., and Morgan, D.S., 2009, Geologic setting and hydrogeologic units of the Columbia Plateau Regional Aquifer System, Washington, Oregon, and Idaho: U.S. Geological Survey Scientific Investigations Map 3088, 1 sheet. (Also available at http://pubs.usgs.gov/sim/3088/.)

Kendall, Carol, and Coplen, T.B., 2001, Distribution of oxygen-18 and deuterium in river waters across the United States: Hydrological Processes, v. 15, p. 1363–1393.

Kinnison, H.B., and Sceva, J.E., 1963, Effects of hydraulic and geologic factors on streamflow of the Yakima River Basin, Washington: U.S. Geological Survey Water-Supply Paper 1595, 134 p., (Also available at http://pubs.er.usgs.gov/usgspubs/wsp/wsp1595.)

Lane, R.C., 2009, Estimated water use in Washington, 2005: U.S. Geological Survey Scientific Investigations Report 2009–5128, 30 p. (Also available at http://pubs.usgs.gov/sir/2009/5128/.)

Lane, R.C., and Whiteman, K.J., 1989, Ground-water levels spring 1985, and ground-water level changes spring 1983 to spring 1985, in three basalt units underlying the Columbia Plateau, Washington and Oregon: U.S. Geological Survey Water-Resources Investigations Report 88–4018, 4 pls. (Also available at http://pubs.er.usgs.gov/usgspubs/wri/wri884018.)

Leek, F., 2006, Hydrogeological characterization of the Palouse Basin Basalt Aquifer System, Washington and Idaho: Pullman, Wash., Washington State University, Master of Science in Engineering thesis, 38 p.

Legates, D.R., and McCabe, G.J., 2005, A re-evaluation of the average annual global water balance: Physical Geography, v. 26, p. 467–479.

Lindolm, G.F., and Vaccaro, J.J., 1988, Region 2, Columbia Lava Plateau, in Back, William, Rosenshein, J.S., and Seaber, P.R., eds., Hydrogeology: Boulder, Colo., Geological Society of America, Geology of North America, v. 0-2, p. 37–50.

Lite, K.E, and Grondin, G.H., 1988, Hydrogeology of the basalt aquifers near Mosier, Oregon—A ground water resource assessment: Oregon Water Resources Department Ground Water Report No. 33, 131 p.

Long, P.E., and Wood, B.J., 1986, Structures, textures, and cooling histories of Columbia River basalt flows: Geological Society of America Bulletin, v. 97, no. 9, p. 1144–1155.

Lum, W.E., II, Smoot, J.L., and Ralston, D.R., 1990, Geohydrology and numerical model analysis of ground-water flow in the Pullman-Moscow area, Washington and Idaho: U.S. Geological Survey Water-Resources Investigations Report 89–4103, 73 p. (Also available at http://pubs.er.usgs.gov/usgspubs/wri/wri894103.)

MacDonald, G.A., 1967, Forms and structures of extrusive basaltic rocks, in Hess, H.H., and Poldervaart, A., eds., Basalts—The Poldervaart treatise on rocks of basaltic composition, v. 1: New York, Wiley, 482 p.

Mather, J.R., 1969, The average annual water balance of the world, in Symposium on Water Balance in North America, 7th Proceedings: Banff, Alberta, Canada, American Water Resources Association, p. 29–40.

McCabe, G.J., and Markstrom, S.L., 2007, A monthly water-balance model driven by a graphical user interface: U.S. Geological Survey Open-File Report 2007–1088, 6 p. (Also available at http://pubs.usgs.gov/of/2007/1088/.)

McCabe, G.J., and Wolock, D.M., 1999, Future snowpack conditions in the western United States derived from general circulation model climate simulations: Journal of the American Water Resources Association, v. 35, p. 1473–1484.

McFarland, W.D., and Morgan, D.S., 1996, A description of the ground-water flow system in the Portland Basin, Oregon and Washington: U.S. Geological Survey Water-Supply Paper 2470–A, 58 p., 7 pls. (Also available at http://pubs.er.usgs.gov/publication/wsp2470A.)

Meier, P.M., Carrera, Jesús, and Sánchez-Vila, Xavier, 1999, A numerical study on the relationship between transmissivity and specific capacity in heterogeneous aquifers: Ground Water, v. 37, no. 4, p. 611–617.

Miller, D.A., and White, R.A., 1998, A conterminous United States multi-layer soil characteristics data set for regional climate and hydrology modeling: Earth Interactions, v. 2., accessed October 1, 2009, at http://EarthInteractions.org.

Montgomery Water Group, 2003, Columbia Basin project water supply, use, and efficiency report: Montgomery Water Group, Inc., 56 p.

Morgan, D.S., and McFarland, W.D., 1996, Simulation analysis of the ground-water flow system in the Portland Basin, Oregon and Washington: U.S. Geological Survey Water-Supply Paper 2470–B, 83 p. (Also available at http://pubs.er.usgs.gov/publication/wsp2470B.)

Myers, C.W., and Price, S.M., 1979, Geologic studies of the Columbia Plateau, a status report: Richland, Wash., Rockwell International, Rockwell Hanford Operations Report RHO-BWI-ST-4, 520 p.

Newcomb, R.C., 1969, Effect of tectonic structure on the occurrence of groundwater in the basalt of the Columbia River Group of the Dalles area, Oregon and Washington: U.S. Geological Survey Professional Paper 383-C, 33 p.

Packard, F.A, Hansen, A.J., Jr., and Bauer, H.H., 1996, Hydrogeology and simulation of flow and the effects of development alternatives on the basalt aquifers of the Horse Heaven Hills, south-central Washington: U.S. Geological Survey Water-Resources Investigations Report 94–4068, 92 p., 2 pls. (Also available at http://pubs.er.usgs.gov/usgspubs/wri/wri944068.)

Poeter, E., 1980, Mathematical models of the gamma-gamma and neutron epithermal neutron geophysical logging processes: Pullman, Wash., Washington State University, Ph.D. dissertation, 184 p.

PRISM Climate Group, 2004, Prism Climate Group: Oregon State University, accessed October 1, 2009, at http://www.prismclimate.org.

PRISM Climate Group, 2010, Latest prism data: Oregon State University, accessed September 3, 2010 at http://www.prismclimate.org.

Prych, E.A., 1983, Numerical simulation of ground-water flow in Lower Satus Creek Basin, Yakima Indian Reservation, Washington: U.S. Geological Survey Water-Resources Investigations Report 82–4065, 78 p. (Also available at http://pubs.er.usgs.gov/usgspubs/wri/wri824065.)

Reidel, S.P., Johnson, V.G., and Spane, F.A., 2002, Natural gas storage in basalt aquifers of the Columbia Basin, Pacific Northwest USA—A guide to site characterization: Richland, Wash., Pacific Northwest National Laboratory, 277 p., accessed June 16, 2011, at http://www.pnl.gov/main/publications/external/technical_reports/PNNL-13962.pdf.

Senay, G.B., 2008, Modeling landscape evapotranspiration by integrating land surface phenology and a water balance algorithm: Algorithms, v. 1, p. 52–68.

Senay, G.B., Budde, M.E., Morgan, D., and Dinicola, R., 2009, Characterizing landscape evapotranspiration dynamics in the Columbia Plateau using remotely sensed data and global weather datasets, *in* Proceedings of the 8th Biennial USGS Pacific Northwest Science Conference, Integrating Science for a Changing Pacific Northwest, Portland, Ore., March 3–5, 2009.

Senay, G.B., Budde, M., Verdin, J.P., and Melesse, A.M., 2007, A coupled remote sensing and simplified surface energy balance approach to estimate actual evapotranspiration from irrigated fields: Sensors, v. 7, p. 979–1000.

Senay, G.B., Verdin, J.P., Lietzow, R., and Melesse, A.M., 2008, Global reference evapotranspiration modeling and evaluation: Journal of American Water Resources Association, v. 44, no. 4, p. 969–979.

Smoot, J.L., and Ralston, D.R., 1987, Hydrogeology and a mathematical model of ground-water flow in the Pullman-Moscow region, Washington and Idaho: Moscow, Idaho, University of Idaho, Idaho Water Resources Research Institute, 118 p.

Snyder, D.T., and Haynes, J.V., 2010, Groundwater conditions during 2009 and changes in groundwater levels from 1984 to 2009, Columbia Plateau Regional Aquifer System, Washington, Oregon, and Idaho: U.S. Geological Survey Scientific Investigations Report 2010–5040, 12 p. (Also available at http://pubs.usgs.gov/sir/2010/5040/.)

Stonestrom, David A., and Harrill, James R., 2007, Ground-water recharge in the arid and semiarid southwestern United States; climatic and geologic framework, in Stonestrom, David A., Constantz, Jim, Ferre, Ty P.A., and Leake, Stanley A., eds., Ground-water recharge in the arid and semiarid southwestern United States: U.S. Geological Survey Professional Paper 1703-A, p. 1-27. (Also available at http://pubs.usgs.gov/pp/pp1703/a/.)

State of Washington Office of Financial Management, 2007, Washington's rank in the nation's agriculture, 2007 Data Book: Office of Financial Management, accessed June 16, 2009, at http://www.ofm.wa.gov/databook/resources/nt14.asp.

Strait, S.R., and Mercer, R.B., 1987, Hydraulic property data from selected test zones on the Hanford Site: Richland, Wash., Rockwell International, Rockwell Hanford Operations, SD-BWI-DP-051, 44 p.

Steinkampf, W.C., and Hearn, P.P., 1996, Ground-water geochemistry of the Columbia Plateau Aquifer System, Washington, Oregon, and Idaho: U.S. Geological Survey Open-File Report 95–467, 67 p. (Also available at http://pubs.er.usgs.gov/usgspubs/ofr/ofr95467.)

Strzepek, K.M., and Yates, D.N., 1997, Climate change impacts on the hydrologic resources of Europe—A simplified continental scale analysis: Climatic Change, v. 36, p. 79–92.

Sublette, W.R., 1986, Rock mechanics data package, Rev. 1: Richland, Wash., Rockwell Hanford Operations Report SD-BWI-DP-041, 78 p.

Swanson, D.A., and Wright, T.L., 1978, Bedrock geology of the southern Columbia Plateau and adjacent areas, Chap. 3, *in* Baker, V.R., and Nummedal, D., eds., The channeled scabland: Washington, D.C., National Aeronautical and Space Administration, Planetary Geology Program, p. 37–57.

Swanson, D.A., Wright, T.L., Hooper, P.R., and Bentley, R.D., 1979, Revision in the stratigraphic nomenclature of the Columbia River Basalt Group: U.S. Geological Survey Bulletin 1457-G, 59 p.

Thornthwaite, C.W., and Mather, J.R., 1955, The water balance. Publications in Climatology, vol.8, Centerton, N.J., C.W. Thornthwaite & Associates, 104 p.

Tolan, T.L., Reidel, S.P., Beeson, M.H., Anderson, J.L., Fecht, K.R., and Swanson, D.A., 1989, Revisions to the estimates of the areal extent and volume of the Columbia River Basalt Group, *in* Reidel, S.P., and Hooper, P.R., eds., Volcanism and tectonism in the Columbia River flood-basalt province: Geological Society of America Special Paper 239, p. 1–20.

Tomkeieff, S.I., 1940, The basalt lavas of the Giant's Causeway district of Northern Ireland: Bulletin Volcanologique, v. 6, p. 89–143.

Trainer, F.W., 1988, Plutonic and metamorphic rocks, *in* Back, W., Rosenshein, J.S., and Seaber, P.R., eds., The geology of North America: Boulder, Colo., Geological Society of America, Hydrogeology, v. O-2, p. 367–380.

U.S. Census Bureau, 2010a, Census 2000 TIGER®/Line Data: U.S. Census Bureau, accessed June 16, 2011, at http://www.census.gov/geo/www/tiger/.

U.S. Census Bureau, 2010b, Population estimates: U.S. Census Bureau, accessed June 16, 2011, at https://www.census.gov/popest/estimates.html.

U.S. Department of Agriculture, 1997, National Engineering Handbook, Part 652, Irrigation Guide: U.S. Department of Agriculture, 820 p.

U.S. Department of Agriculture, 2007a, The Census of Agriculture: U.S. Department of Agriculture, accessed June 16, 2011, at http://www.agcensus.usda.gov/Publications/2007/Online_Highlights/Rankings_of_Market_Value/Washington/index.asp.

U.S. Department of Agriculture, 2007b, 2007 Idaho Cropland Data Layer: U.S. Department of Agriculture, accessed June 16, 2011, at http://datagateway.nrcs.usda.gov/.

U.S. Department of Agriculture, 2007c, 2007 Oregon Cropland Data Layer, accessed June 16, 2011, at http:// datagateway.nrcs.usda.gov/.

U.S. Department of Agriculture, 2007d, 2007 Washington Cropland Data Layer, accessed June 16, 2011, at http:// datagateway.nrcs.usda.gov/.

U.S. Geological Survey, 1975, Water resources of the Toppenish Creek basin, Yakima Indian Reservation, Washington: U.S. Geological Survey Water-Resources Investigations 74–42, 144 p.

U.S. Geological Survey, 2010, Estimated use of water in the United States—County-level data for 1985, 1990, 1995, 2000, and 2005, accessed June 16, 2011, at http://water. usgs.gov/watuse/data/2005/.

Vaccaro, J.J., 1986, Plan of study for the regional aquifer-system analysis, Columbia Plateau, Washington, northern Oregon, and northwestern Idaho: U.S. Geological Survey Water-Resources Investigations Report 85–4151, 25 p.

Vaccaro, J.J., 1999, Summary of the Columbia Plateau Regional Aquifer-System analysis, Washington, Oregon, and Idaho: U.S. Geological Survey Professional Paper 1413-A, 51 p. (Also available at http://pubs.er.usgs.gov/ usgspubs/pp/pp1413A/.)

Vaccaro, J.J., Jones, M.A., Ely, D.M., Keys, M.E., Olsen, T.D., Welch, W.B., and Cox, S.E., 2009, Hydrogeologic framework of the Yakima River Basin Aquifer System, Washington: U.S. Geological Survey Scientific Investigations Report 2009–5152, 106 p. (Also available at http://pubs.usgs.gov/sir/2009/5152/.)

Vaccaro, J.J., and Olsen, T.D., 2007, Estimates of ground-water recharge to the Yakima River Basin aquifer system, Washington, for predevelopment and current land-use and land-cover conditions: U.S. Geological Survey Scientific Investigations Report 2007–5007, 30 p. (Also available at http://pubs.usgs.gov/sir/2007/5007/pdf/sir20075007.pdf.)

Vaccaro, J.J., and Sumioka, S.S., 2006, Estimates of ground-water pumpage from the Yakima Basin Aquifer System, Washington, 1960–2000: U.S. Geological Survey Scientific Investigations Report 2006–5205, 56 p. (Also available at http://pubs.usgs.gov/sir/2006/5205.)

Vermeul, V.R., Cole, C.R., Bergeron, M.P., Thorne, P.D., and Wurstner, S.K., 2001, Transient inverse calibration of site-wide groundwater model to Hanford Operational Impacts from 1943 to 1996—Alternative conceptual model considering interaction with uppermost basalt confined aquifer: Richland, Wash., Pacific Northwest National Laboratory, PNNL-13623 [variously paged].

Wagner, R.J., and Lane, R.C., 1994, Selected ground-water information for the Columbia Plateau Regional Aquifer System, Washington and Oregon, 1982–1985—Volume III, Ground-water quality data: U.S. Geological Survey Open-File Report 93–359, 226 p.

Washington State Department of Health, 2010, Water system data: Washington State Department of Health, accessed June 16, 2011, at http://www.doh.wa.gov/ehp/dw/.

Waters, A.C., 1960, Determining direction of flow in basalts: American Journal of Science, v. 258, p. 350–366.

Whiteman, K.J., 1986, Ground-water levels in three basalt hydrologic units underlying the Columbia Plateau in Washington and Oregon, spring 1984: U.S. Geologic Survey, Water-Resources Investigations Report 86–4046, 4 sheets.

Whiteman, K.J., Vaccaro, J.J., Gonthier, J.B., and Bauer, H.H., 1994, The hydrogeologic framework and geochemistry of the Columbia Plateau Aquifer System, Washington, Oregon, and Idaho: U.S. Geological Survey Professional Paper 1413-B, 73 p.

Wolock, D.M., and McCabe, G.J., 1999, Effects of potential climatic change on annual runoff in the conterminous United States: Journal of the American Water Resources Association, v. 35, p. 1341–1350.

Wood, W.W., and Fernandez, L.A., 1988, Volcanic rocks, Chapter 39, in Back, W., Rosenshein, J.S., and Seaber, P.R., eds., The Geology of North America: Boulder, Col., Geological Society of America, Hydrogeology, v. O-2, p. 353–365.

Woodward, D.G., Gannett, M.W., and Vaccaro, J.J., 1998, Hydrogeologic framework of the Willamette Lowland Aquifer System, Oregon and Washington: U.S. Geological Survey Professional Paper 1424-B, 82 p.

Yates, D.N., 1996, WatBal—An integrated water-balance model for climate impact assessment of river basin runoff: International Journal of Water Resources Development, v. 12, p. 121–140.

Zimmerman, D.A., 1983, Report on task 2 of modeling task force support—Interpret and revise conceptual model—Data documentation from the extended Pasco Basin Model: Richland, Wash., Pacific Northwest Laboratory, 39 p.

Appendix A.   Methods and Assumptions Used to Estimate Lateral Hydraulic Conductivity from Specific-Capacity Data (From Vaccaro and others, 2009)

The modified Theis equation (Ferris and others, 1962, p. 99) was used to estimate transmissivity of the pumped interval for wells with a screened or perforated interval. Transmissivity is the product of hydraulic conductivity and thickness of the part of the hydrogeologic unit supplying water to the well. The modified equation is

$$s = \frac{Q}{4\pi T} \ln \frac{2.25Tt}{r^2 S}, \qquad (A1)$$

where

s is drawdown in the well, in feet;

Q is discharge, or pumping rate, of the well, in cubic feet per day;

T is transmissivity of the hydrogeologic unit, in foot squared per day;

t is length of time the well was pumped, in days;

r is radius of the well, in feet; and

S is storage coefficient, a dimensionless number, assumed to be 0.0001 for basalt and bedrock units and 0.1 for overburden units.

Assumptions for using equation 1 are that aquifers are homogeneous, isotropic, and infinite in extent; wells are fully penetrating; flow to the well is lateral; steady-state drawdown has been achieved; and water is instantaneously released from storage. Additionally, for unconfined aquifers, drawdown is assumed to be small in relation to the saturated thickness of the aquifer. Although many of the assumptions are not precisely met, the field conditions in the study area approximate most of the assumptions and the calculated hydraulic conductivities are reasonable estimates.

A computer program was used to solve equation 1 for transmissivity (T) using Newton's iterative method (Carnahan and others, 1969). The iterative approach is necessary because T cannot be easily solved by a direct-solution method. Next, the following equation was used to calculate hydraulic conductivity:

$$K_h = \frac{T}{b}, \qquad (A2)$$

where

K_h is lateral hydraulic conductivity of the geologic material in the vicinity of the well opening, in feet per day; and

b is thickness, in feet, approximated using the length of the open interval as reported in the driller's report.

The use of the length of a well's open interval for b may overestimate values of K_h because the equations assume that all the water flows laterally within a layer of this thickness. Although some of the flow will be outside this region, the amount can be expected to be small because in most deposits, vertical flow is inhibited by layering. The difference in computed transmissivity, between using 0.1 and 0.0001 for the storage coefficient, is a factor of only about 2. For wells that had no data available for the screen interval or time of pumping, values of 100 feet and 1 hour, respectively, were used.

www.ingramcontent.com/pod-product-compliance
Lightning Source LLC
Chambersburg PA
CBHW081602170526
45166CB00009B/2789